Research, Development, and Technological Innovation

Research, Development, and Technological Innovation

Recent Perspectives on Management

Edited by
Devendra Sahal
International Institute of
Management—Berlin

LexingtonBooks
D.C. Heath and Company
Lexington, Massachusetts
Toronto

Library of Congress Cataloging in Publication Data
Main entry under title:

Research, development, and technological innovation.

Based on papers presented in December 1978 at an international conference on R&D management and research policy held at the International Institute of Management of the Science Center, Berlin.
1. Research—Management. 2. Technological innovations. 3. Technology and state. I. Sahal, Devendra.
Q180.55.M3R47   338.9   79-3095
ISBN 0-669-03377-4

Copyright © 1980 by D.C. Heath and Company

All rights reserved. No part of this publication may be reproduced or transmitted in any form or by any means, electronic or mechanical, including photocopy, recording, or any information storage or retrieval system, without permission in writing from the publisher.

Published simultaneously in Canada.

Printed in the United States of America.

International Standard Book Number: 0-669-03377-4

Library of Congress Catalog Card Number: 79-3095

Contents

	List of Figures	vii
	List of Tables	ix
	Preface	xi
Part I	*R&D Management*	1
Chapter 1	Information Preferences and Attention Patterns in R&D Investment Decisions *Sandra L. Schwartz* and *Ilan Vertinsky*	5
Chapter 2	Research and Development Strategies for Swedish Companies in the Farm Machinery Industry *Harry Nyström* and *Bo Edvardsson*	39
Chapter 3	R&D, Corporate Growth, and Diversification *Ove Granstrand*	53
Chapter 4	Some Behavioral Aspects of the Coupling and Performance of R&D Groups: A Field Study *Richard T. Barth*	91
Chapter 5	Organizational Factors in R&D and Technological Change: Market Failure Considerations *Almarin Phillips*	105
Part II	*Technological Change*	127
Chapter 6	Technical Progress and Evolution *Josef Steindl*	131
Chapter 7	The Scale Factor in Research and Development *Mark F. Cantley* and *Vladimir N. Glagolev*	143
Chapter 8	Diffusion, Innovation, and Market Structure *Stephen Davies*	153
Chapter 9	Technological Progress and Policy *Devendra Sahal*	171

Part III	Science and Technology Policy	199
Chapter 10	**Technology Transfer and Entrepreneurship** *Michael O. Bruun*	203
Chapter 11	**Linking Research Planning to Sectoral Planning** *B. Schwarz*	215
Chapter 12	**Remarks on the Formulation of Technology Strategy** *Karol I. Pelc*	231
Chapter 13	**Scientists on Technology: Magic and English-Language "Industrispeak"** *Michael Fores*	239
Chapter 14	**A Theoretical Basis for Input-Output Analysis of National R&D Policies** *Derek de Solla Price*	251
	Index	261
	About the Contributors	271
	About the Editor	275

List of Figures

3-1	Correlations between Displaced Time Series for Sales and R&D	75
4-1	Path Model of Coupling and Group Effectiveness with Path Coefficients	96
5-1	The Core Organizational Framework	107
5-2	An Extended Horizontal Organizational Framework	108
5-3	An Extended Vertical Organizational Framework	109
6-1	Succession of Steps in Technical Developments Leading to Oxygen-blown Converter Steel	136
7-1	Growth of Market and Scale of Plant	147
7-2	The Scale Growth "Mechanism"	148
8-1	Prediction Generated by the Theoretical Model	155
8-2	Diffusion. (a) Logistic, or Cumulative Normal (b) Cumulative Lognormal	157
8-3	The Level of Diffusion at t and the Size Distribution of Firms	164
9-1	Relationship between Fuel Consumption Efficiency of Tractors and Cumulated Tractor Production, 1920-1968	183
9-2	Relationship between Fuel Consumption Efficiency of Farm Tractors and Average Farm Size, 1920-1968	186
10-1	The Technology Transfer Process	204
10-2	Distribution of Entrepreneurs by Education	207
10-3	Distribution of Entrepreneurs by Father's Occupation	208
10-4	Distribution of Entrepreneurs by Age at Time of Startup	209
11-1	Simplified Outline of Ansoff's Model	221
11-2	Planning Model in the Preliminary Study of the Trp Project	222
11-3	Revised Planning Model in the Trp Project	223

List of Tables

1-1	Variables Relevant for R&D Decision Making	6
1-2	Distribution of R&D Laboratories in Population and in Sample	10
1-3	Executive Characteristics	11
1-4	Firm Characteristics, Statistical Traits	12
1-5	Firm Characteristics, Perceived Role	13
1-6	Factor Analysis Results	16
1-7	Correlations among Factors	18
1-8	Ranking of Items by Average Importance Ratings	19
1-9	Statistical Tests of the Discriminant Analyses on Forty-four Information Items	20
1-10	Results of Discriminant Analysis: Discriminant Dimensions, Location of Group Centroids, and Important Discriminating Variables	22
1-11	Results of Discriminant Analysis for Cross-Classified Groups on Forty-four Economic Items	28
2-1	Relationship between Company Size and R&D Success	44
2-2	Relationship between Company Size and Research Cooperation	46
2-3	Relationship between Research Cooperation and Success in R&D for Companies with Fewer than Nine Employees	47
2-4	Relationship between Research Cooperation and Success in R&D for Companies with Ten to Thirty-nine Employees	48
2-5	Relationship between Research Cooperation and Success in R&D for Companies with More than Forty-nine Employees	49
3-1	Periods in the History of E&S-Corp.	56
3-2	Type and Strategies for Product Diversification Still in Effect in 1975 in the Corporations Studied	59

3-3	Examples of Statistics on Product R&D	65
3-4	Aggregate Figures of Present Situations of the Corporations Studied	66
3-5	Parts of Scott's Model of Stages of Corporate Development	73
4-1	Description of Intergroup Climate Dimensions Significantly Related to Intergroup Communication	94
4-2	Path Contributions to Group Performance Using Path Coefficients	97
8-1	Relative Sizes of Innovators	166
9-1	Parametric Estimates of the Learning-by-Doing Formulation in Various Cases of Technological Innovation	182
9-2	Parametric Estimates of the Specialization-via-Scale Formulation in Various Cases of Technological Innovation	185
9-3	Parametric Estimates of the Learning-via-Diffusion Formulation in Various Cases of Technological Innovation	189
9-4	Long-Term Role of Learning in Various Cases of Technological Innovation	191
9-5	Long-Term Role of Scale of Operations in Various Cases of Technological Innovation	191
9-6	Significance of Differences in the Long-Term Role of Learning in Various Cases of Technological Innovation	192
9-7	Relative Importance of Long-Term Role of Learning and Scale in the Process of Technological Change	192
10-1	A Classification of New Firms	209
10-2	Distribution of the Sample on the Different Branches of Industry	211
12-1	Average Time Horizons for Various Technical Activities	234
12-2	Attributes Affecting the Time Horizon for a Technology Strategy	235

Preface

It is generally agreed that technological change has been and remains a crucial factor affecting long-term economic growth and development. In particular, the growing awareness of scarcity of resources in recent years has led to a tremendous upsurge of interest in the management of technological change. There has been a virtual flood of publications on the various aspects of research and development activity. However, there remain significant gaps in our understanding of the process of technological change.

One major obstacle to a systematic study of technology is the unavailability of the relevant data. Indeed, much of the theoretical literature on technological innovation is bedeviled by the lack of an empirical underpinning. On the other hand, the few theoretical insights that have been unearthed in the empirical investigations are limited in their breadth of coverage. They cannot be readily generalized across institutional backgrounds and time periods. One is left with a lack of both theory and measurement of technology.

This book is an attempt to come to grips with some of the problems in the study of innovative activity. It focuses on three types of problems: the management of research and development at the level of a firm, the process of technological innovation, and sectoral and national technological policy. It reports the results of some of the latest work in these areas that is being done in a variety of disciplines in many different countries.

The initial versions of the chapters in this book were presented in December 1978 at an international conference on R&D management and research policy, sponsored by the Berlin Senator für Wissenschaft und Forschung and held at the International Institute of Management of the Science Center, Berlin. Their publication gives me the welcome opportunity to express my gratitude to Roger Dunbar, Manfred Fleischer, Walter Goldberg, Ove Granstrand, Leah Ireland, and Ilan Vertinsky for their numerous helpful suggestions in this endeavor, and especially to my secretaries, Brigitte Schmidt and Marilyn Shustock, who painstakingly typed many more versions of the manuscript than they care to remember.

Part I
R&D Management

The five chapters in this section are devoted to a single topic: management of research and development activity at the level of the firm. Following the early work of Herbert Simon in organization behavior and the adherents of his school, we recognize that firms generally make their decisions, not through a process of optimization, but rather through selected use of available information. The complexity of real-world problems tends to prevent firms from reaching optimal solutions to their problems in any meaningful sense. Moreover, there exists severe limitations on the extent to which individual decision-making units can process available information. Thus firms often use standard operating procedures and rules of thumb instead of maximizing the value of some objectively specified utility function. This should not be taken to imply any irrationality on the part of decision makers. They act as well as they can while economizing on the necessarily limited human information-processing capability. Such an approach is entirely consistent with what is sometimes termed *procedural* rationality of economic agents. Indeed it can be shown that the use of standard operating procedures corresponds to an optimal strategy in a formal theoretical sense. The nature of these operating procedures has been a subject of considerable inquiry in recent years. However, the thrust of the work to date has been on problems of a very general nature. Little is known about which if any standard operating procedures are used in specific problem areas such as management of research and development activity.

The first chapter, by Schwartz and Vertinsky, investigates this problem: Which information preferences can be said to exist in R&D decision making? According to the results from their analysis of data based on an extensive field survey, R&D decision making is seldom part of the firm's strategy to adapt to its larger environment. The selection of the R&D projects is generally based on a consideration of certain project-specific characteristics such as rate of return, probability of technical success, and expected payback period. In contrast, the indicators of general economic trends and monetary policy are frequently ignored. To this, however, one must add the qualification that the information preferences of small firms differ somewhat from those of big firms. In particular, there seems to exist a size threshold phenomenon in the R&D decision procedures. Regardless of firm size, however, it is perhaps safe to conclude that R&D management is no exception to the heuristic type of decision making.

The relationship between firm size and R&D activity is analyzed in considerable detail in the chapter by Nyström and Edvardsson. The analysis itself is based on an investigation of more than one-hundred Swedish companies in the farm machine industry. Firm size emerges as a significant variable on at least two counts. First, differences in size tend to be systematically related to differences in the types of cooperation between firms engaged in R&D and external organizations such as universities, research institutes, and consultants. The larger companies make greater use of research cooperation than the smaller companies, who rely for the most part on their presumably more limited internal resources. Second, firm size is an important determinant of success in R&D in terms of both the number of new products and the number of patented inventions in relation to the R&D effort in man-years. In particular, the relatively smaller companies tend to be more successful in R&D.

The nature of cooperation between the firms and, say, the universities is also an important variable. Regardless of firm size, temporary and problem-specific cooperation is expected to yield fruitful results in the form of new products. The patentable inventions, on the other hand, seem to require a somewhat more systematic and organized form of cooperation, a strategy particularly suited to the needs of small and medium-sized firms.

The chapters by Granstrand and by Barth present an analysis of the role played by the external marketing and internal organizational factors, respectively, in the management of R&D. The focal point of Granstrand's investigation is on the relationship between the R&D activities and corporate development, particularly corporate growth and diversification. This relationship has many facets, and the results from the several hundred interviews with employees of the major Swedish multinational corporations reported in this study help clarify many of the hitherto unresolved issues. There are four principal findings of the study based on an analysis of both historical and interview data. First, there is considerable evidence of "grass-root R&D" underlying innovative output. This type of R&D is devoted to a seemingly mundane type of activity. However, in a cumulative sense its role in the process of innovation may be rather significant. Second, the relationship between R&D and corporate development tends to be mutually causal. There is considerable give-and-take and mutual reinforcement between the two. Feedback between R&D and corporate growth is often characterized by a time lag as well as by contemporary coupling through budgetory allocations. In magnitude, the time lag between R&D and the degree of diversification tends to be greater than the time lag between R&D and sales growth. This is to be expected, insofar as success in diversification requires that R&D-generated sales in a new product must grow not only in an absolute sense but also in relation to sales in other product areas. Third, while the initial phases of innovative activity are generally a consequence of the demand-pull mechanism, the later phases of it seem to be due to technology push. Very often a new product (or process) has its origin in the existence of a certain demand for it. Subsequently, however, it is not uncommon to find that the range of applicability of the product is extended far beyond the initially perceived demand for it simply because there exist purely technical opportunities to do so. Fourth, while many of the innovations underlying corporate diversification originate from within the firm, the others originate from without. That is, there is little support for the commonly held belief that technological change of an incremental nature is a consequence of R&D activity within the established firms while radical innovations generally come from without. Success in R&D-based diversification therefore depends on organizational permeability, that is, the susceptibility of the firm to external ideas as well as to the internal impulses to go beyond the existing product areas.

Barth analyzes the role played by three key elements in task-dependent R&D units: intergroup climate, communication, and conflict resolution. Three meaningful dimensions of intergroup climate emerge from the analysis of the data from a field study: cohesion between the team members, risk taking, and clarity. They are significantly related to the intergroup communication. Communication, in turn, has a significant impact on the modes of conflict resolution as well as on R&D success in terms of reduced uncertainty in technical problem solving and in meeting the target dates of specific projects. In summary, innovations depend to a considerable extent on effectiveness of communication within the R&D laboratories.

Phillips presents a unified theory of R&D activity based on a consideration of both internal organizational and external marketing factors within a single analytical framework. Of special interest is the function of R&D in various intra- and interorganizational systems. These systems are in turn composed of subsystems concerned with selling, manufacturing, and R&D operations. The last are further decomposable into exploratory research, advanced development, engineering, and product and marketing activities. Two principal conclusions emerge from a consideration of the interdependent nature of these organizations. First, nowhere is the heuristic nature of problem solving more evident than in R&D management. Second, vertical integration of firms is far more conducive to R&D and technological innovation than is the market mechanism of relevant transactions.

1 Information Preferences and Attention Patterns in R&D Investment Decisions

Sandra L. Schwartz and
Ilan Vertinsky

Simon and others have argued that firms make decisions on the basis of partial knowledge (March and Simon 1958; Cyert and March 1963). High costs associated with obtaining and processing information, constrained computational abilities, and limited spans of attention induce the development of search heuristics. These heuristics are often institutionalized in the form of standard operating procedures for the selection and evaluation of alternatives. The importance of information preference and attention patterns for R&D policies is often ignored. The problem of R&D project selection constitutes one of the more complex areas for firm decision making. It deals with risky alternatives, whose outcomes are realized over a long time horizon and are subjected to risks from both internal and environmental sources. Generally, the focus of R&D inducement policies has been on creating economic incentives or direct investment in R&D. Attempts to employ information strategies to induce R&D seem to ignore the fact that what is good news to some is no news to others. Identifying differences in information attended to makes it possible to design strategies and interventions to create favorable climates for R&D for particular target populations, thereby increasing their potential impact.

In this study we focus on empirical investigation of information preferences and attention patterns in R&D decision making. In particular, we investigate the association of preference and attention patterns to decision makers' attributes.

Methodology: The Experiment

An "information basket" was constructed, combining dimensions judged relevant to R&D decision making according to a variety of normative perspectives and behavioral theories. Table 1-1 lists the items and their sources in the

---

This study was supported by funds from the Department of Industry, Trade, and Commerce, Canada. It is also a part of an international comparative study supported by funds from NATO. The funding agencies are not responsible for its content. The authors would like to thank V.F. Mitchell and J.W.C. Tomlinson for helpful discussions as this study progressed and T.A. Cameron and A. Lansberg for computer assistance.

Table 1-1
Variables Relevant for R&D Decision Making

Concept	Item	Source
Economy, General		
Past and current trends	Average profit rate in the economy	Keynes (1964), Galbraith (1973), Preston (1975), Trendicator (Baguley and Booth, 1975)
	Short-term bank interest rates[a]	
	Stock market trends[a]	
	General trends in inventories	
	General trends of growth	
	Unemployment	
Demand changes	Expected growth of real GNP	Schumpeter (1971)
	Growth of population	Quinn (1966), Bright (1970)
Inducements to change	Expected wage settlements	Schmookler (1966), Rosenberg (1974), Kamien and Schwartz (1968), Fellner (1971), Hayami and Ruttan (1970)
	Expected general productivity changes	
	Expected rate of inflation	
	Expected energy requirements	
Demand and cost changes	Expectations with respect to the foreign exchange rate	Leonard (1971)
Economy, Government Role		
Direct costs	Government subsidies for R&D projects[a]	Tilton (1971), Brooks (1972), Quinn (1966), Hamberg (1966), Leonard (1971), Foster (1971), Arrow (1962)
	Low-interest government loans for R&D projects[a]	
	Possibility of gaining a new government contract for part of the project[a]	
	Favorable tax policies for R&D projects	

Category	Item	Reference
Indirect costs	Accelerated depreciation of R&D expenditures for tax purposes	Hamberg (1966)
	Accelerated depreciation of new capital equipment expenditures for tax purposes[a]	
	Low-interest rates on government bonds	Keynes (1964), Galbraith (1973)
	High-interest rates on government bonds	
	Interest rates on government bonds increasing	
	Interest rate on government bonds declining	
Information	Government funding of feasibility studies for R&D projects[a]	Thurston (1971), Brooks (1972), Bright (1970), Foster (1971), Drucker (1975)
	Availability of sound government information on technological change[a]	
	Government support and promotion for market development	
	Availability of government surveys of market potential[a]	
Market influence	Growth of government expenditures	Bright (1970), Foster (1971), Thurston (1971), Smith (1973)
	Favorable tariff policy	
	Pollution control measures (environmental concern)	
Information		
Private	Availability of private surveys of market potential	Thurston (1971), Brooks (1972), Bright (1970), Foster (1971), Drucker (1975)
Government	Availability of government surveys of market potential[a]	
	Availability of sound government information on technological change[a]	
	Government funding of feasibility studies for R&D projects[a]	

Table 1-1 *(continued)*

Concept	Item	Source
Market	Average profit rate of industry group Stability of market Barriers to entry in the market	Ansoff and Stewart (1967) Mansfield (1969), Schumpeter (1971), Scherer (1971), Comanor (1967), Williamson (1965), Baldwin & Childs (1969), Cooper (1966), Galbraith (1973)
Firm Demand	Recent growth of sales of firm Expected growth of sales of firm Stage in life cycle of existing products	Lithwick (1969), Leonard (1971), Mansfield (1968), Hamberg (1966) Kotler (1967), Quinn (1966), Tilles (1966), Ansoff and Stewart (1967)
Supply Factors	Availability of scientifically trained personnel	Brooks (1972), Cooper (1966)
Liquidity	Average profit rate of firm Accelerated depreciation of new capital equipment expenditures for tax purposes[a] Short-term bank interest rates[a] Stock market trends[a] Government subsidies for R&D projects[a] Low-interest government loans for R&D projects[a]	Scherer (1971), Williamson (1965), Galbraith (1973), Cyert and March (1963), Mansfield (1968), Hamberg (1966), Tilton (1971), Brooks (1972)
Innovativeness	History of success with R&D (firm's)	Ansoff and Stewart (1967)
Project Commitment of money, and resources	Cost of the R&D project relative to total sales of firm Possibility of gaining a new government contract for part of the project[a]	Mansfield (1964), Scherer (1971), Gerstenfeld (1970), Cooper (1966), Tilles (1966), Ansoff and Stewart (1967), Mottley and Newton (1959), Allen (1970) Brooks (1972), Tilton (1971), Hamberg (1966)

Information Preferences and Attention Patterns

Commitment of time	Expected payback period for the R&D project	Leonard (1971), Mansfield (1968), Gerstenfeld (1970), Bright (1970), Kotler (1967), Ansoff and Stewart (1967), Brooks (1972), Tilles (1966), Cooper (1966), Allen (1970), Hamberg (1963)
Profitability	Rate of return for the R&D project	Mansfield (1969), Disman (1962), Quinn (1966), Kotler (1967), Peterson (1967), Allen (1970)
	Expected impact of the R&D project on market share	Mansfield (1968), Peterson (1967), Mottley and Newton (1959)
	Expected change in sales attributed to R&D project	Peterson (1967)
Risk	Probability of technical success estimated for the R&D project	Scherer (1971), Mansfield (1968), Nelson (1959), Gerstenfeld (1970), Disman (1962), Quinn (1966), Ansoff and Stewart (1967), McGlauchlin (1968), Thurston (1971), Allen (1970), Cooper (1966), Tilles (1966), Cranston (1974), Mottley and Newton (1959)
	Patentibility of innovation	Phillips (1966), Scherer (1971), Ansoff (1965), Mansfield (1968), Quinn (1966), Foster (1971)

aItem appears in more than one dimension.

literature classified into six general dimensions: general economic trends and expectations, government role in the economy, information, market, firm, and project attributes. This information basket was used as a basis for constructing a questionnaire to solicit executive preferences for information items. Subjects were asked to rate the items on a seven-point Likert scale. The questionnaire is presented in Appendix 1A. Analysis of these responses using a variety of multivariate techniques identified relationships between preferences and individual and organizational attributes.

A sample of 330 executives (R&D directors and presidents) was randomly selected from the directory of R&D establishments in Canada (Ministry of State for Science and Technology 1974) plus the remainder of "Top 100" firms (Morgan 1975) not included in the R&D directory.[1] As shown in table 1-2 the response distribution corresponds well with the population distribution, indicating that the sectoral breakdown in the sample is representative of R&D establishment frequencies in the population.

The questionnaire solicited supplemental information relating to the attributes of the executive and his firm. The distributions of these attributes for the respondents are summarized in table 1-3. Included in the sample are fifteen corporate presidents. Sixty-one percent of the respondents are currently em-

Table 1-2
Distribution of R&D Laboratories in Population and in Sample
(in percents)

Industry	Population	Sample
Agriculture	0.9	1.5
Forestry	0.7	0.8
Mines and wells	5.8	5.3
Food and beverages	5.8	6.8
Tobacco	0.4	–
Rubber	5.5	4.5
Leather	0.4	–
Textiles	1.0	0.8
Wood	0.6	0.8
Furniture	0.2	–
Paper	3.0	4.5
Printing	0.2	–
Primary metals	2.8	3.8
Metal fabricating	5.6	5.3
Machinery	11.5	8.3
Transport equipment	5.0	3.0
Electrical products	16.0	13.6
Nonmetallic minerals	2.4	2.3
Petroleum and coal	1.9	3.0
Chemicals	18.0	22.0
Miscellaneous manufacturing	4.5	3.8
Construction	0.4	1.5
Transportation and communication	2.2	4.5
Services	4.5	3.8

Table 1-3
Executive Characteristics

Category Label	Code	Absolute Frequency	Relative Frequency (percent)
Position as given by respondent			
Not given	0	9	6.8
President	1	15	11.4
Vice-president, controller	2	18	13.6
Chief enginer, chemist	3	5	3.8
Director	4	23	17.4
Assistant director	5	2	1.5
Manager, superintendent	6	51	38.6
Supervisor, head	7	3	2.3
Operations planner, advisor	8	3	2.3
Research associate, etc.	9	3	2.3
Total		132	100.0
Age			
Not given	0	6	4.5
20-25	1	2	1.5
25-30	2	7	5.3
30-40	3	33	25.0
40-50	4	42	31.8
50+	5	42	31.8
Total		132	100.0
Present department			
Not given	0	8	6.1
Production	1	3	2.3
Sales, marketing, advertising	2	6	4.5
Finance, accounting	3	4	3.0
Research, development	6	81	61.4
General administration	7	22	16.7
Other	8	8	6.1
Total		132	100.0
Formal education			
Not given	0	6	4.5
High school	1	4	3.0
Certificate degree	2	2	1.5
Bachelor's degree	3	35	26.5
Postgraduate training	4	25	18.9
Master's degree	5	35	26.5
Doctoral degree	6	25	18.9
Total		132	100.0

ployed in the R&D departments of their firms. The general level of formal education is quite high: two-thirds of the executives have had at least some postgraduate training.

The characteristics of the firms are divided into statistical traits and role perception. Table 1-4 displays the statistical characteristics. Half of the sample firms have sales over $50 million; half are either privately owned or controlled by a few interests. The sample is equally split in terms of Canadian or foreign control and represents well the state of ownership in Canadian manufacturing.

For an indication of the perceived role of the firm, respondents were asked to rate on a seven-point Likert scale market type (stable-volatile), innovation

Table 1-4
Firm Characteristics, Statistical Traits

Category Label	Code	Absolute Frequency	Relative Frequency (percent)
Sales			
Not given	0	5	3.8
0-999	1	13	9.8
1,000-9,999	2	30	22.7
10,000-49,999	3	25	18.9
50,000-74,999	4	8	6.1
75,000-99,999	5	4	3.0
100,000-250,000	6	13	9.8
250,000+	7	34	25.8
Total		132	100.0
Owned			
Not given	0	8	6.1
Public, widely held	1	33	25.0
Public, controlled by few	2	48	36.4
Privately owned	3	43	32.4
Total		132	100.0
Employees			
Not given	0	12	9.1
1-49	1	14	10.6
50-99	2	11	8.3
100-249	3	14	10.0
250-499	4	16	12.1
500-749	5	8	6.1
750-999	6	4	3.0
1,000-1,499	7	8	6.1
1,500-2,999	8	15	11.4
3,000-4,999	9	4	3.0
5,000+	10	26	19.7
Total		132	100.0
Control			
Not given	0	9	6.8
Yes	1	60	45.5
No	2	63	47.7
Total		132	100.0

(follower-innovator), and importance of R&D (no involvement-extremely important). Table 1-5 presents the distributions of the results. Approximately 50 percent considered their market somewhat stable while 12 percent considered their market extremely volatile. Few considered their firm followers. R&D involvement was extremely important for 22 percent.

Confirming the A Priori Dimensions of the Information Basket and Testing its Relevance

Factor analysis was used to confirm whether the a priori dimensions correspond to empirical patterning of variables. Both general theory and observation of the

Table 1-5
Firm Characteristics, Perceived Role

Category Label	Code	Absolute Frequency	Relative Frequency (percent)
Market Type (Stable=1, volatile=7)			
	0	6	4.5
	1	11	8.3
	2	26	19.7
	3	26	19.7
	4	27	20.5
	5	20	15.2
	6	10	7.6
	7	6	4.5
Total		132	100.0
Firm (Follower=1, innovator=7)			
	0	4	3.0
	1	1	0.8
	2	4	3.0
	3	14	10.6
	4	22	16.7
	5	41	31.1
	6	25	18.9
	7	21	15.9
Total		132	100.0
R&D import (No Involvement=1, extreme=7)			
	0	5	3.8
	2	4	3.0
	3	16	12.1
	4	21	15.9
	5	28	21.2
	6	29	22.0
	7	29	22.0
Total		132	100.0

scree line indicated that six to eight factors could be reasonably extracted. Factor analysis was therefore run, with varimax rotation being constrained to six, seven, and eight factors, respectively. Analysis of this factor range indicated that seven factors yielded the optimal granularity level. Results of the varimax rotation indicated that some of the items had high loadings on a number of factors defining correlated factors. Therefore, direct oblimin was performed for a somewhat correlated factor space (delta = -0.05).[2] The resulting factors were (1) general economic conditions, (2) profitability and sales, (3) government support for R&D, (4) project evaluation, (5) information, (6) taxes, and (7) inducements for R&D.

Table 1-6 presents the significant factor loadings with items ordered according to their a priori theoretical clusters (see appendix 1B for abbreviations of items). The relative weights for each factor were calculated as $\sum_i l_{ij}^2 / \sum_i \sum_j l_{ij}^2$ where l_{ij} is the factor pattern loading of item i on variable j for the variables loading significantly on each factor. The weights are presented in the bottom line of table 1-6. The factor space explains about 60 percent of the variation of the items.

Analysis of interfactor correlations revealed the following associations. The tax dimension correlates positively with general economic trends and more weakly with government support for R&D. Profitability and sales correlate positively with project evaluation (both dimensions include R&D project evaluation items). Information correlates negatively with both the profitability and inducement factors. Government support and information are also somewhat positively correlated, while government support and taxes are somewhat negatively correlated. (The correlations among factors are reported in table 1-7.)

A few items load significantly on two factors, but analysis reveals that such duality is theoretically warranted. For example, general trends in growth loaded on both the factor representing the dimension of general economic conditions and the factor representing the profitability and sales dimension. Expected inflation (loading on factors 1 and 6) affects the formation of general economic expectations and appears as a tax or reduction of real profits. Favorable tax policies (loading on factors 3 and 6) appears as a tool of government support for R&D and as a reduction in tax. Expected growth of government expenditures (loading on factors 1 and 3) affects formation of expectations and appears as government support for R&D, apparently reflecting the assumption that growth of government expenditures means more contracts and grants for R&D and market support.

Two items are not significant in the factor space: average profit of the industry group and availability of scientifically trained personnel.

The factor analysis confirmed to a large extent the hypothesized decision dimensions (table 1-2). Government items separate into five groups clustering with other items depending on function. Interest on government bonds and growth of government expenditures cluster with the general economic items to form factor 1, general economic conditions. Government surveys and government information cluster with private surveys to form factor 5, information.

Tax-related items cluster with inflation to form factor 6, taxes. Pollution control and favorable tariffs cluster with expected productivity change and energy requirements to form factor 7, inducements to R&D. The remaining government items form factor 3, government support of R&D.

The market characteristics and project characteristics also split up. Factor 2, sales and profitability, includes firm characteristics, market stability, and project-related items that reflect change in sales and risk. Barriers to entry (as a risk item) and patentability cluster with project commitment items (relative size and payback) and the rate of return.

Information Preferences

General Patterns

The most important items are those pertaining to the following attribute dimensions: the *project*, the *firm*, and the *market*. Most of the high-ranking environmental items were project specific (with expected growth of sales for the firm completing the set of important decision cues). The items receiving the lowest ratings included indicators of general economic trends and monetary policies (see table 1-8 for rankings). The details of item rankings for each attribute dimension follow.

General Economic. The most important general economic items are general trends of growth and expected general productivity changes. Each was given an importance rating of 5 or more by approximately 50 percent. Expected rate of inflation, expected real GNP growth, and average profit rate in the economy are bimodally distributed and are important for some of the respondents.

Government. Government subsidies and favorable tax policies for R&D received high ratings from at least 60 percent of the subjects. The possibility of receiving government contracts, which also diminishes the firm's financial commitment to R&D, was bimodally distributed; it received high importance ratings from 35 percent and low ratings from 50 percent of the subjects.

Accelerated depreciation of R&D expenditures was also important for 45 percent of the respondents, while accelerated depreciation of capital equipment has no special importance for R&D decision making.

Interest rates on government bonds have little importance to R&D decisions, supporting Galbraith's thesis that R&D projects are internally funded or that costs are passed on. Originally it was hypothesized that the effect of government bond interest rates on R&D decision making would vary with the

Table 1-6
Factor Analysis Results

	General Economic Conditions	Profitability and Sales	Government Support	Project Evaluation	Information	Taxes	Inducements
I. Economy, General							
A. Trends							
AV FR ECON	.44						
SH T BANK I	.51						
STK MKT TR	.61						
INVENT TR	.32						
GEN TRD GR	.42	.47					
UNEMPLMT	.57						
B. Future expectations							
EXP GNP GR	.48						
POP GR	.46						
C. Cost changes (inducements)							
EX W SETTL	.51						
FOR EXCH R	.44						
EXP INFL	.46					.44	
EXP PRD CH							.48
EXP E REQ							.56
II. Economy, Government role							
A. Direct costs							
GOVT SUBS			−.62				
L I GOVT L			−.38				
P GOVT CON			−.72				
FAV TX POL			−.39			.60	
B. Indirect costs							
AC DEPR RD						.79	
ACCL DEPRE						.72	
I I GOVT B	.54						
C. Market influence							
GOV EXP GR	.40		−.45				
FAV TARIF							.33
POLL CONT							.53
D. Information and support							
GOV SUP MK			−.66				
FEAS STUD			−.80				
GOVT SURV					−.73		
GOV INF CH					−.60		
III. Other information							
PRIV SURV					−.70		
IV. Market							
AV PR GRP							
STABL MKT		.35					
BAR TO ENT				.40			

Table 1-6 *(continued)*

	General Economic Conditions	Profit-ability and Sales	Government Support	Project Evaluation	Information	Taxes	Inducements
V. Firm							
A. Demand							
REC SLS GR		.47					
EXP SLS GR		.73					
LIFE CYCL		.47					
B. Supply, Factors							
SCI TR PRS							
C. Liquidity							
AV PR FIRM		.59					
D. Innovativeness							
RD HISTRY		.30					
VI. Project							
A. Commitment							
COST SLS				.56			
EXP PAY B				.87			
B. Profitability							
RD ROR				.67			
RD MKT SHR		.59					
RD SALE CH		.59					
C. Risk							
P TECH SUC		.36					
PATENTS		.32		.33			
Relative Weights	.22	.20	.18	.13	.10	.11	.07

level and direction of change. For this reason four interest items were included. However, given the low importance ratings of all and the similarity of their distributions, it was apparent that they all measured the same thing. Therefore only one interest item (increasing interest on government bonds) was retained for further analysis.

Government funding of feasibility studies is important to 40 percent of the sample. Other forms of government information all rank low.

Pollution control measures are important to 40 percent of the sample, perhaps reflecting low concern with the environmental impact of changing technologies.

Information. Though no information item receives a high importance rating, private information sources are generally more important than governmental.

Market. At least 50 percent of the respondents rate stability of market and barriers to entry as important. This probably indicates the requirement of low risk levels or the comfort of concentration for R&D investment decisions.

Table 1-7
Correlations among Factors

	General Economic Conditions	Profitability and Sales	Government Support	Project Evaluation	Information	Taxes	Inducement
General economic conditions	1.00	0.14	−0.16	0.04	−0.24	0.30	0.23
Profitability and sales		1.00	−0.20	0.31	−0.33	0.12	0.22
Government support			1.00	−0.20	0.29	−0.27	−0.09
Project evaluation				1.00	−0.23	0.14	0.16
Information					1.00	−0.26	−0.32
Taxes						1.00	0.20
Inducement							1.00

Firm. The expected future is much more important than the past in making R&D decisions except with respect to the R&D history of the firm. Expected growth of sales of the firm is quite important, rated high by 70 percent of the subjects, while recent growth of sales was rated high by 50 percent. The firm's history of past success with R&D is given a high rating by 50 percent of the subjects.

The profit rate of the firm is important for 60 percent of the respondents, while outside sources of funds (except government subsidies) are all low (again indicating preference for internal funding).

Life cycle considerations have high importance ratings or greater than average importance for 55 percent of the subjects.

Project. The most important project attributes are the probability of success (rated high by 90 percent) and the rate of return (80 percent). These are closely followed by expected payback period and change in sales atrributed to the project (80 percent high). Other important project attributes were the expected impact on market share (70 percent high), and the cost of the project relative to sales (60 percent high), and patentability (50 percent).

Discriminating among Groups

Discriminant function analysis was employed to examine differences in information preferences and relate them to executive attributes and firm character-

Table 1-8
Ranking of Items by Average Importance Ratings

	Rank	Item	Mean Score	Rank	Item	Mean Score	
	1.	RD ROR	5.72	25.	ACCL DEPR	3.39	
	2.	P TECH SUC	5.64	26.	EXP E REQ	3.33	
Extremely* high	3.	EXP PAY B	5.42	27.	GOV SUP MK	3.27	
	4.	RD SALE CH	5.35	28.	EXP GNP GR	3.13	
	5.	RD MKT SHR	5.05	29.	FAV TARIF	3.08	
	6.	EXP SLS GR	4.99	30.	EXP INFL	3.05	
	7.	COST SLS	4.84	31.	AV PR ECON	2.97	
	8.	AV PR FIRM	4.72	32.	P GOVT CON	2.94	
	9.	SCI TR PRS	4.53	33.	EX W SETTL	2.89	
	10.	BAR TO ENT	4.42	34.	GOV INF CH	2.67	
	11.	GEN TRD GR	4.32	35.	GOVT SURV	2.56	
	12.	PATENTS	4.32	36.	L I GOVT L	2.49	
	13.	FAV TX POL	4.32	37.	POP GR	2.38	
	14.	RD HISTRY	4.23	38.	SH T BANK I	2.32	
	15.	STABL MKT	4.22	39.	GOV EXP GR	2.08	
	16.	GOVT SUBS	4.17	40.	FOR EXCH R	2.01	
	17.	EXP PRD CH	4.04	41.	INVENT TR	1.91	
	18.	LIFE CYCL	4.02	42.	UNEMPLMT	1.77	
	19.	REC SLS GR	3.87	43.	STK MKT TR	1.72	Extremely low
	20.	AC DEPR RD	3.83	44.	I I GOVT B	1.39	
	21.	POLL CONT	3.54	45.	H I GOVT B	1.36	
	22.	PRIV SURV	3.43	46.	D I GOVT B	1.27	
	23.	FEAS STUD	3.42	47.	L I GOVT B	1.23	
	24.	AV PR GRP	3.39				

Note: *Each mean score of the seven highest items is significantly higher ($\alpha = 0.05$) than each mean score of the seven lowest items.

istics. Using the levels of each attribute and some combinations of attributes, we defined alternative classifications of subjects. The ratings of the information items were used as potential independent variables in the various discriminant functions. The percentage of the respondents correctly classified is a measure of the adequacy of the analysis. Half of the groupings yielded accurate predictions for at least 80 percent of the cases (table 1-9). The discriminant analysis performed better than the probability of maximum allocation in all cases. On average the discriminant analysis performed 30 percentage points better than the probability of allocation by chance, indicating a high degree of accuracy. The canonical correlations for each function are also presented in table 1-9.[3]

The important discriminating variables for each classification of subjects are presented in table 1-10 along with the group centroids. When there are only two groups, the group centroids are separated by at least one standard deviation. Important variables are defined to be those with standardized coefficients of 0.4 and greater. The dimensions are defined by the factors associated with the important variables. In the case of two functions, some variables are important for both; varimax rotation would be a useful tool to give better separation of the dimensions, but this option was not available.

Determination of the group that gives highest rating to the items helps in

Table 1-9
Statistical Tests of the Discriminant Analyses on Forty-four Information Items

Groupings Variables	Definition of Groups	Percentage Correct	Chance Probability	Maximum Probability	Canonical Correlation[a]
Position	Presidents Other management Staff	94	66	80	.71, .56
Department	R&D Others	75	54	65	.55
Age	<40 ≥40	73	56	67	.52
Education	Bachelor's degree Postgraduate	82	59	71	.56
Location	Ontario Quebec	84	58	70	.65
Industry	High R&D (> 3 percent value added) Medium R&D (1 - 3 percent value added) Low R&D (< 1 percent value added)	87	47	52	.74, .62
Market	Stable (1-3) Volatile (4-7)	79	50	50	.57
Firm	Follower (1-5) Leader (6,7)	77	54	64	.54
R&D importance	Little involvement (1-4) Important (5-7)	83	56	68	.65
Sales	< $1 million $1-50 million > $50 million	71	41	46	.70, .40

Employment	< 100 employees	75	36	44	.72, .53
	100-1000 employees				
	> 1000 employees				
Owned	Public, widely held	64	35	39	.55, .43
	Public, control by few				
	Private				
Canadian	100 percent ownership	81	40	51	.68, .60
	50 percent-99 percent ownership				
	< 50 percent ownership				
Control	Yes	82	50	51	.62
	No				

[a]In order of importance when there is more than one discriminant function.

Table 1-10
Results of Discriminant Analysis: Discriminant Dimensions, Location of Group Centroids, and Important Discriminating Variables

		Location of Group Centroid		*Order of Important Variables and Group Attention*			
Grouping Variables	*Name of Dimension*	*Function 1*	*Function 2*	*Function 1*		*Function 2*	
Position	General economic conditions, information	1. 1.59	−0.81	SH T BANK I	3,1,2	GOVT SURV	3,2,1
		2. −0.10	0.26	GOV INF CH	3,1,2	BAR TO ENT	3,2,1
		3. −1.57	−1.55	INVENT TR	3,2,1	RD ROR	2,1,3
				FOR EXCH R	3,2,1	EXP PAY B	3,2,1
	Government support, project evaluation			I I GOVT B	3,2,1	GOV SUP MK	3,1,2
				EX W SETTL	1,3,2		
Department	General economic conditions, government support, sales	1. 0.40		GOVT SURV	2,1		
		2. −0.75		EX W SETTL	2,1		
				RD MKT SHR	1,2		
				FAV TX POL	2,1		
				EXP SLS GR	2,1		
				GOV INF CH	1,2		
				SH T BANK I	2,1		
Age	General economic conditions, tax, sales	1. 0.73		AV PR ECON	2,1		
		2. −0.37		AV PR FIRM	1,2		
				FAV TX POL	2,1		
				ACCL DPR	1,2		
				UNEMPLMT	2,1		
				I I GOVT B	1,2		
Education	General economic conditions, sales, project evaluation, information	1. 0.86		FEAS STUD	1,2		
		2. −0.36		SHT BANK I	2,1		
				COST SLS	2,1		
				EXP GNP GR	2,1		
				EXP SLS GR	2,1		

Information Preferences and Attention Patterns

Location	General economic conditions, information, government support	1. −0.43 2. 0.99		SH T BNK I GOV INF CH INVENT TR GOVT SURV P GOVT CON EX W SETTL	1,2 1,2 2,1 2,1 1,2 2,1	SCI TR PRS[a] PATENTS[a] INVENT TR[a] P GOVT CON[a] EXP E REQ	1,2,3 3,1,2 1,3,2 2,3,1 2,1,3
Industry	Information, taxes, inducement, availability of resources	1. 0.74 2. −0.46 3. −2.29	0.32 −0.45 2.44	AC DEPR RD GOV INF CH SCI TR PRS POLL CONT	3,2,1 3,2,1 1,2,3 2,1,3		
	General economic conditions, risk, government support, inducements, availability of resources						
Market	General economic conditions, government support, and project evaluation	1. 0.57 2. −0.57		I I GOVT B FOR EXCH R FAV TX POL FEAS STUD RD ROR INVENT TR EXP PAY B	1,2 2,1 2,1 1,2 1,2 1,2 1,2		
Firm	General economic conditions, sales, government support	1. −0.41 2. 0.72		P TECH SUC EXP SLS GR GOV SUP MK L I GOVT L EXP GNP GR	2,1 2,1 2,1 1,2 2,1		
R&D importance	Government support, sales	1. 0.94 2. −0.45		GOV SUP MK RD HISTRY	2,1 2,1		
Sales	General economic condition, sales	1. −1.51 2. −0.35 3. 0.66	−0.81 0.41 −0.20	SH T BANK I EXP SLS GR I I GOVT B LIFE CYCL	2,1,3 1,2,3 3,2,1 1,2,3	SH T BANK I FOR EXCH R P GOVT CON LIFE CYCL	2,1,3 1,3,2 1,2,3 1,3,2
	General economic conditions, sales, government support						

Table 1-10 *(continued)*

Grouping Variables	Name of Dimension	Location of Group Centroid		Order of Important Variables and Group Attention	
		Function 1	Function 2	Function 1	Function 2
Employment	General economic conditions, inducement, sales	1. −1.15 2. −0.22 3. 0.72	0.58 −0.70 0.28	SH T BANK I 1,2,3 INVENT TR 3,2,1 POLL CONT 3,2,1 POP GR 3,2,1 REC SLS GR 1,2,3	PRIV SURV 2,3,1 AC DEPR RD 3,1,2 FAV TX POL 2,1,3 L I GOVT L 1,2,3 REC SLS GR 1,2,3 P GOVT CON 1,3,2
	Tax, government support, sales, information				
Owned	General economic conditions	1. −0.08 2. −0.58 3. 0.71	−0.70 0.30 0.21	SH T BANK I 3,1,2 INVENT TR 2,1,3 STK MK TR 1,3,2 FOR EXCH R 3,1,2 UNEMPLMT 1,3,2	UNEMPLMT 1,3,2 FAV TARIF 2,3,1 I I GOVT B 3,2,1 LIFE CYCL 1,3,2
	General economic conditions, sales, and inducements				
Canadian	General economic conditions, government support, sales	1. −0.97 2. 0.32 3. 0.52	−0.09 1.32 −0.38	FAV TX POL 1,3,2 I I GOVT B 1,3,2 COST SLS 1,3,2 EXP GNP GR 1,3,2 GOV SUP MK 1,3,2	BAR TO ENT 1,3,2 FAV TX POL 1,3,2 AC DEPR RD 3,1,2 EX GNP GR 1,3,2 RD HISTRY 3,1,2
	General economic conditions, sales, project evaluation, government support, tax				
Control	General economic conditions, government support, tax	1. −0.63 2. 0.60		FAV TX POL 1,2 EX GNP GR 2,1 SH T BANK I 1,2 I I GOVT B 2,1 AC DEPR RD 1,2	

[a]Importance ratings <0.4.

interpreting the group distinctions. The discussion of results that follows highlights the relationship of differences in information preferences to decision makers' attributes and to organizational characteristics and individual attribute interactions.

Executive Attributes and Information Preferences

Generally, a preference for a broader basket of information for R&D project evaluation was indicated by the following overlapping groups: (1) the young, (2) the executives with longer time of schooling, and (3) executives perceiving their firms as leader rather than follower.

Presidents focused relatively more than other executives on evaluation of general growth trends and wage settlements, that is, general expectations of investment climates. Senior managers seemed to focus on information related to expected returns on specific proposals and availability of government support. Other executives (staff) were less discriminating, indicating a broader set of variables as important to R&D decision making. This result confirms the expectation that top-level executives concern themselves with strategic problems, while the next echelon of senior management focuses on tactical problems.

Analysis of the departmental membership of subjects yields support to the observation that information selection reflects parochial tendencies. R&D managers, in comparison to others, focus more on information about technical change, while executives in other departments display interest in general market information. The exception to this rule is the high interest of R&D managers in assessment of the impact of R&D on market share as opposed to general investment climates and contribution to overall company growth.

Executives based in Quebec tend more than Ontario executives to value information concerning inventory trends, government support for market development, and expected wage settlements, while the latter pay more attention to conditions of short-term financing and direct government demand for R&D services. These distinctions were judged to be a function of different industrial structures in the two provinces.

Organizational Attributes and Information Preferences

The analysis of discriminating information preferences among subject classifications by organizational attributes first focused on the firm's commitment to R&D. Firms with high R&D commitment displayed a relatively low interest in information about inducements and incentives for R&D. This is perhaps a reflection of the institutionalization of R&D investments as a standard operating procedure in the perceived general market role of these organizations. Low commitment to R&D was associated with higher concern with tax incentives and patentability of developments. Medium R&D involvement was associated with preference for information about derived demands for R&D. For example, the probability of government contracts and expected energy requirements received higher ratings by executives in firms with medium R&D involvement.

Executives in firms with highly stable markets showed higher preferences than other executives for information on general economic trends, reflecting their concern with a longer planning horizon. Those in volatile markets tended to have higher preferences for short-term information, such as foreign exchange rates and existing tax incentives.

Executives who perceived their firms as leaders, tended to pay more attention to general economic conditions than those perceiving their firms to be followers. The latter showed more interest in positive inducements for R&D involvement.

The effects of sales size on information preferences of executives seemed to confirm Galbraith's claim that large firms are concerned with government bond interest as a measure of general economic conditions. Executives from medium and small firms, in contrast, paid more attention to bank interest rates which affect their costs.

Discriminant functions derived to predict classification of executives by the employment size of their firms indicated the following preferences. Executives from smaller firms were more attentive to information items concerned with liquidity and direct support for R&D. The distance in group centroids between executives from medium and large firms was less than the distance between either of them and executives from small firms. This may reflect a size threshold phenomenon in organizational decision procedures.

Ownership and firm control have several interesting associations with information behavior of executives. Executives from publicly owned firms whose shares are widely held tended to pay more attention than other executives to stock market trends, unemployment levels, and product life cycle characteristics. Those from privately owned firms showed more concern than others with interest rates and foreign exchange rates. These differences may reflect the alternative potential financial sources perceived by these groups for R&D investment funding as well as differences in performance evaluation criteria.

The three groups of executives classified by the degree of Canadian control in their firms were equally distant in the discriminant space. Both dimensions included general economic conditions, government support, and sales. The second dimension included also project evaluation and tax. The first-dimension items were all rated as highly important by group 1 (100 percent Canadian ownership) followed by group 3 (less than 50 percent ownership). This dimension separates group 1 from the other groups. The second dimension separates group 2 from groups 1 and 3. Group 3 was most concerned with accelerated depreciation of R&D and R&D history, while group 1 rated higher the remaining items in the second dimension. Where Canadian ownership was sufficient for control, the executives were concerned with favorable tax policies, short-term bank interest, and accelerated depreciation of R&D. In firms not controlled by Canadian interests, executives were concerned with general economic conditions such as expected growth of GNP and the interest on government bonds.

Interactions of Firm and Executive Attributes

To identify possible impacts of interactions of individual and organizational attributes on information behavior, individual attributes (position and department) were cross-classified with firm attributes (industry, market, firm R&D importance, sales, employment, ownership, and Canadian control) to identify new groupings of subjects. To keep the number of groups manageable, we defined only two position groups: presidents and other management. Table 1-11 summarizes the results of the analysis. The discriminant function always performed better than both the probability of chance allocation and the probability of maximum allocation. The highlights of the analysis are described for the two major discriminating dimensions of the groupings.

Position and Location. Dimension 1 included government bonds, government information, favorable tax policy, inventory trends, and short-term bank interest. Dimension 2 included expected wage settlements, sales change attributed to the R&D project, pollution control measures, and unemployment. Presidents were similar on dimension 1, which explained 61 percent of the variation. Presidents of Ontario firms were similar to managers from Quebec firms on dimension 2, explaining 22 percent of the variation.

Position and Industry. Dimension 1 included government contracts, government support of markets, government feasibility studies, average profit of the firm, foreign exchange rates, pollution control, inventory trends, population growth, and short-term bank interest. Dimension 2 included accelerated depreciation and government contracts. Presidents in firms with medium and low R&D commitment and managers in firms of high and medium R&D commitment were similar on dimension 1 (explaining 35 percent of the variation). Presidents and managers of medium commitment firms were similar on dimension 2 (30 percent of the variation).

Position and Market. Dimension 1 included government bonds, government information, pollution control, and probability of technical success of the

Table 1-11
Results of Discriminant Analysis for Cross-Classified Groups on Forty-four Economic Items

Grouping Variables	Number of Groups	Percentage Correct	Probability of Chance Allocation	Probability of Maximum Allocation
Position and				
Location	4	83	44	60
Industry	6	84	37	44
Market	4	79	39	47
Firm	4	79	41	57
R&D importance	4	81	43	56
Sales	6	78	33	43
Employment	6	77	31	40
Ownership	6	67	25	34
Control	4	73	39	50
Department and				
Location	4	79	32	47
Industry	6	77	26	33
Market	4	61	28	38
Firm	4	66	30	45
R&D importance	4	69	31	42
Sales	6	67	25	37
Employment	6	71	22	34
Ownership	6	58	19	28
Control	4	72	28	38
Location and				
Industry	6	72	27	37
Market	4	82	29	38
Firm	4	67	32	47
R&D importance	4	79	33	47
Sales	6	63	24	33
Ownership	6	67	20	27
Control	4	79	29	37

project. Dimension 2 included government bonds, sales change attributed to the R&D project, average profit of the firm, foreign exchange rate, inventory trends, and short-term bank interest. Dimension 1 separated presidents in volatile markets from other presidents and all the managers (explaining 51 percent of the variation). On dimension 2 presidents and managers in volatile markets were similar.

Position and Sales. Dimension 1 included government bonds, inventory trends and short-term bank interest. Dimension 2 included general trends in growth, expected wage settlements, expected sales growth, average profit of the industry group, short-term bank interest, and the relative size of the project. For the largest firms position was unimportant. For the smallest firms the presidents and other managers were far apart in the discriminant space. On dimension 1 management in small and medium-sized firms were close. For the medium-sized firms the distance between presidents and management was moderate.

Position and Canadian Control. Dimension 1 included government bonds, private surveys, expected wage settlements, and short-term bank interest. Dimension 2 included government information, government support of markets, average profit of the firm, inventory trends, population growth, short-term bank interest, and the rate of return of the project. Managers were similar whether the firm was Canadian controlled or not. However, presidents of each group differed. For Canadian-controlled firms presidents and management were similar on dimension 2 (explaining 29 percent of the variation). In the non-Canadian group presidents were similar to managers on dimension 1 (explaining 59 percent of the variation).

Department and Market. Dimension 1 included expected productivity change, change in market share attributed to the R&D project, average profit in the economy, expected wage settlements, patents, favorable tax policies, and average profit of the group. Dimension 2 included accelerated depreciation, change in market share attributed to the R&D project, expected wage settlements, expected and recent sales growth, and rate of return of the R&D project. On dimension 1 R&D managers were similar whether the markets were volatile or stable (explaining 46 percent of the variation). However, on dimension 2 R&D managers in stable markets were different from the cluster of all the other groups (explaining 36 percent of the variation).

Department and Employment. Dimension 1 included expected sales growth, pollution control, favorable tax policies, inventory trends, and short-term bank interest. Dimension 2 included government loans, general trends in growth, change in market share attributed to the R&D project, average profit of the group, and growth of government expenditures. Employment size of the firm was more important than department in determining similarity of items attended. R&D and other executives of large firms were similar on both dimensions. On dimension 1 a cluster formed of both executive groups in medium-sized firms and another cluster of executives in small firms (explaining 41 percent of the variation).

Location and Sales. Dimension 1 included accelerated depreciation, government contracts, expected sales growth, and population growth. Dimension 2 included government information, availability of scientifically trained personnel, change in market share attributed to the R&D project, expected wage settlements, and expected sales growth. Sales size was the basis of clusters on the first dimension rather than location (explaining 38 percent of the variation). On the second dimension Ontario firms of all sizes and large Quebec firms formed another similar group (26 percent of the variation).

Conclusions and Policy Implications

The study leads to two general behavioral propositions. First, though the economic environment creates the setting in which R&D decisions are made, project-specific attributes are universally ranked higher than environmental ones. R&D decision making is thus stimulated by the opportunity of particular R&D projects rather than being part of an integral environmental adaptation strategy. Second, there is no marked interest in government participation through direct funding or other assistance programs. This is perhaps indicative of the widely held belief that "there is no free lunch."

In addition, significant differences in information selection patterns among executives have been identified. These were associated with differences in executive and firm attributes. For example, executives low in the hierarchy show parochial tendencies, while top management is more environmentally attuned. In sectors that are volatile the focus is on short-term information; in more stable environments long-term general economic trends are more important. Firms in which R&D is institutionalized pay relatively little attention to inducements, while those whose commitment to R&D is low show greater concern for inducements.

The study suggests that strategies aimed at improvements in the specific attributes of investment opportunities should be universally attended to. In contrast, measures aimed at improving specific climate attributes or measures that provide specific inducements for R&D stimulation should have a highly selective impact. "Social marketing" strategies to stimulate R&D must provide a fit in content to prevailing information selection patterns of executives and organizations. Similarity groupings such as those identified by this study constitute the appropriate target populations for specific strategic designs. Clearly, it is also necessary to ensure that other characteristics of the information diffusion process provide a fit with search and evaluation procedures in firms (in media type, form of messages, and the like) and that barriers to actions be removed. Further studies to provide this information are necessary for improvement in the impact of intervention on R&D investment.

Notes

1. In addition a pilot study with fifty subjects was undertaken to pretest the questionnaire.

2. See Rummel (1970) for a discussion of the extraction of significant factors and rotation of factors to obtain meaningful patterning of the variables.

3. See Morrison (1969) and Mosteller and Bush (1954) for the theory involved in selecting variables to enter the discriminant functions.

References

Allen, J.M. 1970. A survey into the R&D evaluation and control procedures currently used in industry. *Journal of Industrial Economics* 18:161-181.

Ansoff, H.I. 1965. *Corporate strategy: An analytical approach to business policy for growth and expansion.* New York: McGraw-Hill.

Ansoff, H.I., and Stewart, J.M. 1967. Strategies for a technology-based business. *Harvard Business Review* 45:71-83.

Arrow, K. 1962. Economic welfare and the allocation of resources for invention in N.B.E.R. In *The rate and direction of inventive activity.* Princeton, N.J.: Princeton University Press, pp. 609-625.

Baguley, R.W., and Booth, J.J. 1975. *Royal Bank Trendicator Report* 2, no. 8. Economics Department. Toronto: Royal Bank of Canada.

Baldwin, W.L., and Childs, G.L. 1969. The fast second and rivalry in research and development. *Southern Economic Journal* 36:18-24.

Bright, J.R. 1970. Evaluating signals of technological change. *Harvard Business Review* 48:62-70.

Brooks, H. 1972. What's happening to the U.S. lead in technology? *Harvard Business Review* 50:110-118.

Comanor, W.S. 1967. Market structure, product differentiation, and industrial research. *Quarterly Journal of Economics* 81:639-657.

Cooper, A.C. 1966. Small companies can pioneer new products. *Harvard Business Review* 44:162-179.

Cranston, R.W. 1974. First experience with a ranking method for portfolio selection in applied research. *IEEE Transactions on Engineering Management* EM-21:148-152.

Cyert, R.M., and March, J.G. 1963. *A behavioral theory of the firm* Englewood Cliffs, N.J.: Prentice-Hall.

Disman, S. 1962. Selecting R&D projects for profit. *Chemical Engineering* 69:87-90.

Drucker, P.F. 1975. Quoted by R. Perry. Drucker formula, information is power. *Financial Post,* April 5.

Fellner, W. 1971. Empirical support for induced innovation. *Quarterly Journal of Economics* 85:580-605.

Foster, R.N. 1971. Organize for technology transfer. *Harvard Business Review* 49:110-120.

Frank, R.E., Massey, W.F., and Morrison, D.G. 1965. Bias in multiple discriminant analysis. *Journal of Marketing Research* 2:250-258.

Galbraith, J.K. 1973. *Economics and the public purpose.* Boston: Houghton Mifflin.

Gerstenfeld, A. 1970. *Effective management of research and development.* Reading, Mass.: Addison-Wesley.

Hamberg, D. 1966. *R&D: Essays on the economics of research and development.* New York: Random House.

Hayami, Y., and Ruttan, V.W. 1970. Factor prices and technical change in agricultural development: The United States and Japan, 1880-1960. *Journal of Political Economy* 78:1125-1135.

Kamien, M.I., and Schwartz, N.L. 1968. Optimal "induced" technical change. *Econometrica* 36:1-17.

Keynes, J.M. 1964. *The general theory of employment, interest, and money.* New York: Harcourt, Brace and World.

Kotler, P. 1967. Operations research in marketing. *Harvard Business Review* 45:30-44.

Leonard, W. N. 1971. Research and development in industrial growth. *Journal of Political Economy* 79:232-255.

Lithwick, N.H. 1969. *Canada's science policy and the economy.* London: Methuen Publications.

Mansfield, E. 1964. Industrial research and development expenditure: determinants, prospects, and relation to size of firm and inventive output. *Journal of Political Economy* 72.

Mansfield, E. 1969. Industrial research and development: characteristics, costs and diffusion of results. *American Economic Review, Papers and Proceedings* 59:65-71.

March, J.G., and Simon, H.A. 1958. *Organizations.* New York: Wiley.

McGlauchlin, L.D. 1968. Long-range technical planning. *Harvard Business Review* 46:54-64.

Ministry of State for Science and Technology. 1973. *Directory of research and development establishments in Canadian industry.* Ottawa: Information Canada.

Morgan, P. 1975. Inflation helps raise the top 100 to new heights. *Financial Post,* July 26.

Morrison, D.G. 1969. On the interpretation of discriminant analysis. *Journal of Marketing Research* 6:156-163.

Mosteller, F., and Bush, R.R. 1954. Selective quantitative techniques. In A. Lindsey, ed., *Handbook of social psychology,* vol. 1. Reading, Mass.: Addison-Wesley.

Mottley, C.M., and Newton, R.D. 1959. The selection of projects for industrial research. *Operations Research* 7:740-751.

Nelson, R.R. 1959. The simple economics of basic scientific research. *Journal of Political Economy* 67:297-306.

Peterson, R.W. 1967. New venture management in a large company. *Harvard Business Review* 45:68-76.

Phillips, A. 1966. Patents, potential competition, and technical progress. *American Economic Review, Papers and Proceedings* 56:301-310.

Preston, R.S. 1975. The Wharton long-term model: Input-output within the context of a macro forecasting model. *International Economic Review* 16:3-19.

Quinn, J.B. 1966. Technological competition: Europe vs. U.S. *Harvard Business Review* 44:113-130.

Rosenberg, N. 1974. Science, invention, and economic growth. *Economic Journal* 84:90-108.

Rummel, R.J. 1970. *Applied factor analysis.* Evanston, Ill.: Northwestern University Press.

Scherer, F.M. 1971. *Industrial market structure and economic performance.* Chicago: Rand McNally and Company.

Schmookler, J. 1966. *Invention and economic growth.* Cambridge, Mass.: Harvard University Press.

Schumpeter, J. 1971. The instability of capitalism. Reprinted in N. Rosenberg, ed., *The Economics of technological change.* Middlesex: Penguin Books.

Smith, V.K. 1973. A review of models of technological change with reference to the role of environmental resources. *Socio-Economic Planning Sciences* 7:489-509.

Thurston, R.H. 1971. Make TF serve corporate planning. *Harvard Business Review* 49:98-102.

Tilles, S. 1966. Strategies for allocating funds. *Harvard Business Review* 44:72-80.

Tilton, J.E. 1971. *International diffusion of technology: the case of semiconductors.* Washington, D.C.: Brookings Institution.

Williamson, O.E. 1965. Innovation and market structure. *Journal of Political Economy* 73:67-73.

Appendix 1A

Sample Questionnaire on General Economic Environment Factors

Rate the importance of various factors in R&D project evaluation. Please use the rating values from a seven-point scale, where 1 is the minimum (unimportant) and 7 is the maximum (critical). Enter the rating on the line to the left of each fact. If *totally* irrelevant enter 0.

(6) ___ Expected rate of inflation
(7) ___ Accelerated depreciation of new capital equipment expenditures for tax purposes
(8) ___ Possibility of gaining a new government contract for part of the project
(9) ___ History of success with R&D (firm's)
(10) ___ Low interest rates on government bonds increasing
(11) ___ Low-interest government loans for R&D projects
(12) ___ Interest rates on government bonds increasing
(13) ___ General trends of growth
(14) ___ Availability of sound government information on technological change
(15) ___ Availability of private surveys of market potential
(16) ___ Expected payback period for the R&D project
(17) ___ Expected energy requirements
(18) ___ Expected general productivity changes
(19) ___ Availability of scientifically trained personnel
(20) ___ Barriers to entry in the market
(21) ___ Availability of government surveys of market potential
(22) ___ Interest rate on government bonds declining
(23) ___ Expected impact of the R&D project on market share
(24) ___ Government support and promotion for market development
(25) ___ Expected growth of real GNP
(26) ___ Average profit rate in the economy
(27) ___ Expected wage settlements
(28) ___ Expected change in sales attributed to R&D project
(29) ___ Government funding of feasibility studies for R&D projects
(30) ___ Average profit rate of firm
(31) ___ Patentibility of innovation
(32) ___ Expected growth of sales of firm
(33) ___ Expectations with respect to the foreign exchange rate
(34) ___ Pollution control measures (environmental concern)
(35) ___ Stage in life cycle of existing products
(36) ___ Accelerated depreciation of R&D expenditures for tax purposes
(37) ___ Probability of technical success estimated for the R&D project
(38) ___ Favorable tax policies for R&D projects
(39) ___ Average profit rate of industry group
(40) ___ Favorable tariff policy
(41) ___ Growth of government expenditures
(42) ___ Recent growth of sales of firm
(43) ___ Stock market trends
(44) ___ Unemployment
(45) ___ General trends in inventories
(46) ___ High interest rates on government bonds
(47) ___ Growth of population
(48) ___ Government subsidies for R&D projects
(49) ___ Stability of market
(50) ___ Short-term bank interest rates
(51) ___ Cost of the R&D project relative to total sales of firm
(52) ___ Rate of return for the R&D project

Appendix 1B
Glossary of Item
Abbreviations

ACCL DEPR	Accelerated depreciation of new capital equipment expenditures for tax purposes
AC DEPR RD	Accelerated depreciation of R&D expenditures for tax purposes
AV PR ECON	Average profit rate in the economy
AV PR FIRM	Average profit rate of firm
AV PR GRP	Average profit rate of industry group
BAR TO ENT	Barriers to entry in the market
COST SLS	Cost of the R&D project relative to total sales of firm
D I GOVT B	Interest rate on government bonds declining
EXP E REQ	Expected energy requirements
EXP GNP GR	Expected growth of real GNP
EXP INFL	Expected rate of inflation
EXP PAY B	Expected payback period for the R&D project
EXP PRD CH	Expected general productivity changes
EXP SLS GR	Expected growth of sales of firm
EX W SETTL	Expected wage settlements
FAV TARIF	Favorable tariff policy
FAV TX POL	Favorable tax policies for R&D projects
FEAS STUD	Government funding of feasibility studies for R&D projects
FOR EXCH R	Expectations with respect to the foreign exchange rate
GEN TRD GR	General trends of growth
GOV EXP GR	Growth of government expenditures
GOV INF CH	Availability of sound government information on technological change
GOV SUP MK	Government support and promotion for market development
GOVT SUBS	Government subsidies for R&D projects
GOVT SURV	Availability of government surveys of market potential
H I GOVT B	High interest rates on government bonds
I I GOVT B	Interest rates on government bonds increasing
INVENT TR	General trends in inventories
LIFE CYCL	Stage in life cycle of existing products
L I GOVT B	Low interest rates on government bonds
L I GOVT L	Low-interest government loans for R&D projects
PATENTS	Patentability of innovation
P GOVT CON	Possibility of gaining a new government contract for part of the project

POLL CONT	Pollution control measures (environmental concern)
POP GR	Growth of population
PRIV SURV	Availability of private surveys of market potential
P TECH SUC	Probability of technical success estimated for the R&D project
REC SLS GR	Recent growth of sales of firm
RD HISTRY	History of success with R&D (firm's)
RD MKT SHR	Expected impact of the R&D project on market share
RD ROR	Rate of return for the R&D project
RD SALE CH	Expected change in sales attributed to R&D project
SCI TR PRS	Availability of scientifically trained personnel
SH T BANK I	Short-term bank interest rates
STABL MKT	Stability of market
STK MKT TR	Stock market trends
UNEMPLMT	Unemployment

2
Research and Development Strategies for Swedish Companies in the Farm Machinery Industry

Harry Nystrom and
Bo Edvardsson

Farm machines are of increasing importance in the agricultural sector. During this century mechanization has rapidly been introduced into many areas of farm production. In Sweden this has particularly been the case during the 1950s and 1960s. In this period the price relationship between manual labor and machines, together with a shortage of labor and a concentration to larger-sized farms, has made investment in farm machinery highly attractive to most farmers. Since 1950, the cost per hour for manual labor in Sweden has increased seventeen times, while the machine cost per hour has only tripled (Nilsson 1978).

At the same time the market share of Swedish-made farm machines has radically decreased since World War II (Holmström 1976). In 1938, for instance, Swedish tractors completely dominated their home market (93 percent of delivery value), while in 1950 the market share of imported machines was more than 50 percent, and in 1975 it was 72 percent. With regard to other farm machines the market share of imported machines was about 10 percent before World War II and increased to 51 percent in 1975.

What, then, can be the reason for this negative trend in the sales of Swedish-made farm machines on their home market? Part of the explanation may well be that Swedish manufacturers have been less successful in their research and development than their foreign competitors.

This study is an attempt to investigate and evaluate the extent and direction of R&D and product development for Swedish companies making and selling farm machines. The focus is on how successful different companies have been in their R&D effort during the past ten years.

In a broader context this chapter may be viewed as part of an ongoing research project, concerned with research and development strategies for Swedish industrial firms. Prior to the present investigation eleven companies in four lines of business (pharmaceuticals, steel, electronics, and chemicals) were analyzed in an attempt to relate their R&D efforts to their success in developing new products (Nyström 1977a, b, 1978a, b). As a continuation of these studies industrial firms working in the agricultural area have been chosen as a further area of investigation. This chapter, dealing with companies that manufacture farm machines, is the first result of this extension to agribusiness; subsequent studies of other companies in the agricultural area, for instance food companies, have been initiated.

Strategies for research and development so far represent a highly neglected research area in business studies. At the same time rapid technological development, together with other changes in society, makes it necessary for most companies to be highly active in research and development. The organization and direction of R&D in companies is therefore an important area of strategic decision making, and there is a need to develop and test a conceptual framework for dealing with this type of problem.

In Nyström's study of R&D strategies two major factors appeared to be related to success in product development. The first was the external organization of research and development, and the second was the combining and recombining of technologies to achieve new products.

Companies emhasizing outside contacts in their R&D, that is with an external orientation, were more successful in developing new products with a high level of technological innovation than companies that relied predominantly on internal resources and competence for R&D.

Synergistic use of technology, that is, combining different technologies such as biochemistry and immunology, was also associated in this study with a larger number of products and a higher level of technological innovation than isolated technology use, that is, working mainly within established technology areas such as biochemistry or immunology.

The focus in these studies was on technological success as a prerequisite for commercial success. The reason for this approach is that research and development strategies are more directly related to technological success than to marketing measures of success, which depend not only on success in R&D but also on marketing variables such as price and advertising.

If the company is successful in developing technologically qualified products, with high market potential, it will also have a favorable setting for achieving marketing success. Research and development strategies in our approach are thus viewed as setting the stage for the successful marketing of new products.

Empirical support for this approach is to be found in our early studies where other, more marketing-oriented measures of R&D success proved less successful in discriminating between and accounting for the outcome of R&D strategies for the companies studied than our technological outcome measure, the level of technological innovation (Nyström 1977a).

In this study of farm machines we shall attempt to determine how successful the companies have been in their research and development strategies in a ten-year perspective, from 1968 to 1977. We shall use the same basic overall framework that we used in our earlier studies of companies working in other technology and market areas. The wider, more extensive data base and somewhat different problem emphasis and methods used for analyzing the data make this study an independent attempt to test the overall framework more than a direct follow-up of the results of the earlier studies.

Empirical Background

This study of farm machines is concerned with 140 companies manufacturing and selling farm machines in Sweden. Farm machines in this context are defined as equipment for soil preparation, sowing, harvesting, and agricultural transports, together with barn equipment, such as milking machines, manure-handling systems, and grain driers. Forestry and garden machines are not included. The basis for the selection of companies was a list of firms manufacturing and selling farm machines compiled by the Swedish Agricultural Board in 1976. Companies established after 1976 are therefore not included.

The investigation was carried out as a total study of the industry. Data with regard to the organization, direction, and intensity of R&D were collected for 139 of the 140 companies. Only one company declined to take part in the study. For some of the companies farm-machines were only a small part of the product assortment, but for most it was of dominating importance. In the former case only data on R&D for the farm machine products were collected. Since in most of the diversified companies farm machines composed a division or small profit unit with independent R&D, it did not prove too difficult in these instances to separately analyze R&D for the farm machine products.

The data were collected mainly by telephone interviews. Ten companies wanted to give written replies, and in these cases mail inquiry was used. The basis for the questions was an interview guide with open responses. The advantage of this procedure, compared with more fixed questions and answers, is that it makes possible better communication and understanding between interviewer and respondents. For relatively new areas of research such as this one this type of approach permits classifications and reconceptualizations. At the same time a high a degree of comparability between companies was sought, by focusing the questions in the same crucial areas for all companies concerned.

Data on present company size and R&D efforts during the ten-year period 1968-1977 were collected. Both size and R&D efforts were measured by the equivalent of full-time employees on an annual basis, with part-time efforts being added together to give a total estimate in man-years. This measure was judged to be the easiest to respond to by companies; at the same time it provided a satisfactory basis for separating R&D from other activities and for indicating the relative magnitudes of R&D in the different companies.

Comparability rather than financial measures of R&D was thus stressed, since the main aim of the investigation was to establish how successful companies were in relation to each other rather than to achieve profitability estimates for R&D in individual companies. Most companies lacked separate budgets for R&D, and it was therefore almost impossible for them to state R&D expenditures in monetary terms. On the other hand, they usually had a much better knowledge of how many people were working on R&D and to what extent these people spent their time on this type of work. R&D was defined as research effort to develop new products. Technical marketing and production services to aid in the manufacture and sales of products already on the market were excluded. Our definition of a new product requires that it be new from both a technical and a marketing point of view. This means that it must be based on a new technical solution and must satisfy customer demand in a new and better way than existing products do. It was not necessary, however, for the company itself to have originated the idea for the new product, but we required it to have actively participated in the technical and market development of the product.

To give us information about the extent to which companies had outside cooperation in R&D, we asked what environmental contacts the companies had utilized in their product development. Such contacts could be with universities, other research institutes, consultants, other companies, or customers. Finally, we asked the companies whether they had been able to achieve patent protection for their new products.

Analytical Approach

As we have noted, our analysis is based on information from 140 Swedish companies that manufacture and sell farm machines. During the period of investigation, 1968-1977, these companies had developed and marketed 166 new products. With our definition of a new product, 46 companies had not developed any new products during this period.

Two measures of R&D success are employed in our study. The first is the number of new products developed, and the second is the number of new products for which companies have gained patent protection. According to our first, quantitative criteria, companies have been successful if they have achieved a large number of new products. According to our second, more qualitative criteria, a large number of patented products is a further indication of success.

The underlying idea is that patent protection is as good an indication as possible of the level of technological innovation. The more a new product is based on new technical principles and technical solutions not previously applied to the area of application we are concerned with, the higher the level of technological innovation, and the more successful a company has been in its product development according to this criterion.

Ideally we would have liked to directly estimate this level of technological innovation, as in previous work (Nyström 1977a). To do this, however, requires a very detailed and interrogatory method for obtaining data, and for this reason personal interviews were used in our early work.

Since this study is more extensive and was based on telephone interviews, it was not possible to directly estimate the level of technological innovation. Instead, patent information was used as the best possible indicator of the level of technological innovation. Patent laws reflect a similar view of what constitutes a new and therefore patentable product. Therefore the higher the level of technological innovation for a new product, the higher the chances for achieving patent protection should be. Our earlier data also indicate, as might be expected, that the correlation between level of technological innovation and patent protection, while not perfect, is sufficiently high to justify the use of patent protection as an indicator of the level of technological innovation, if other data are not available.

To estimate R&D success it is not enough to consider only the new product outcome. We also need to relate this to R&D effort, to achieve a measure of effectiveness. As our estimate of resource input we have chosen the time spent on R&D by company employees, converting full- and part-time efforts to man-year equivalents. For each company we can then calculate both the average number of new products and the average number of new patented products, for the ten-year period, and relate this to R&D effort. We may then use these two measures to compare the relative success of different companies and R&D strategies in generating new products.

One main result of our previous study was that external contacts and research cooperation to utilize knowledge and resources outside the company in many instances appeared to be a major determinant of R&D success. The other major determinant appeared to be the synergistic use of technology, combining and recombining technologies to achieve breakthroughs in new areas of product development. In the farm machinery sector it soon became evident that the differences between companies were related more to the type and degree of research cooperation than to technology use, and we therefore decided to concentrate on this strategic variable.

The other variable that we chose to emphasize was company size, since both our own earlier work and other studies indicate that there may be fundamental differences in both R&D strategies and R&D success between small and large companies.

Company Size and R&D Success

The relationship between company size and R&D success is a highly controversial issue both in general economic debate and in the literature on innovation (Galbraith 1952; Jewkes, Sawers, and Stillerman 1958; Scherer 1964; Lindström 1972).

Among the arguments in favor of greater success for small companies are statements that such companies provide a more stimulating research environment, are more receptive to new ideas, and can more rapidly develop these into new products. In the case of large companies bureaucratic disadvantages are usually assumed to exist, but these may be counteracted by greater resources for R&D and a greater willingness to take risks.

It is difficult, however, to find empirical evidence for increasing returns to scale in R&D based on company size. Lindström (1971, p. 131) summarizes a review of empirical results pertaining to the relationships between company size and R&D success, stating that it is highly doubtful whether any general size-related advantages exist.

In our study companies were divided into three sizes, based on their total number of employees. Official Swedish company statistics were used as a basis for this classification. A reduction in the number of size groups from six in the official data to three in our calculations was carried out, however, since preliminary checks showed there would be little loss of information. We found the same tendencies with three as with six groups, and the smaller number of groups was therefore chosen to facilitate the analysis and interpretation of the data. In table 2-1 we see that the distribution of companies with regard to size is relatively even, with 38 percent in the small group, 34 percent in the middle group, and 28 percent in the large group. If we look at the total number of new products per year, it is not surprising that the largest companies have been most successful during the ten-year period. If, on the other hand, we view the total number of new products in relation to R&D effort, the results are quite different. The smallest companies have thirteen times as many new products per year, when we divide by the number of man-years devoted to R&D, compared with the largest companies (0.26 compared with 0.02).

Table 2-1
Relationship between Company Size and R&D Success

Variable	Company Size (Total Number of Employees 1977)		
	Up to 9	10-49	50+
Number of companies	53	48	39
(percent)	(38)	(34)	(28)
Total number of new products per year and company during the ten-year period	0.08	0.12	0.17
Total number of patented new products per year and company during the ten-year period	0.02	0.04	0.09
Average number of new products per year and company in relation to R&D effort (number of R&D employees)	0.26	0.15	0.01
Average number of patented new products per year and company in relation to R&D effort (number of R&D employees)	0.075	0.05	0.01

Turning to a comparison between the largest and the middle-sized companies, we again find that the larger companies have a greater absolute number of new products (0.17 compared with 0.12). In this instance, too, the number of new products per year in relation to R&D effort is much smaller for the larger companies (0.02 compared with 0.15).

When comparing the middle-sized companies with the smallest companies, we find similar size-related differences. The middle-sized companies have a greater number of new products than the smallest ones (0.12 compared with 0.08). The smallest companies, on the other hand, have a larger number of new products in relation to R&D effort (0.26 compared with 0.15).

We may conclude, then, that smaller company size in our data is consistently linked to a larger number of new products in relation to R&D effort. Larger company size, as may be expected, is associated with a greater absolute number of products, when we do not consider R&D effort.

If we turn from a more quantitative to a more qualitative discussion of R&D success, our focus now is on the number of patented new products rather than on all new products. Again, larger companies have a larger absolute number of patented new products, when we do not consider R&D effort, than smaller companies (0.07 compared with 0.05 and 0.01). If we consider R&D effort, however, the smaller companies, as in our quantitative analysis of success, again have a greater number of new products than the larger companies (0.07 compared with 0.05 and 0.01).

We may thus summarize our discussion of the relationship between company size and R&D success by stating that the smaller companies in our data have consistently been much more successful than the larger companies. This is true both with regard to our quantitative measure of success—total number of new products in relation to R&D effort—and with regard to our qualitative measure—number of patented products in relation to effort.

External R&D Cooperation and R&D Success

Companies are grouped into three categories, based on their relationships with the R&D environment outside the company. The first group consists of companies that reported little or no external cooperation in R&D during the ten-year period. The second group is made up of companies that reported having taken part in spontaneous and temporary R&D cooperation. The third group consists of companies claiming to have had a systematic and organized cooperation with the outside research community in carrying out R&D.

Before looking directly at the relationships between external research cooperation and success in R&D, we need to investigate size-related differences in such cooperation between companies. As we have shown, there are consistent size-related differences in R&D success between companies in our data.

When analyzing the success of research cooperation, we therefore need to allow for size differences, particularly if these differences are related to differences in research cooperation. We shall do this by carrying out the analysis within each size group of companies and then making comparisons between size groups to see whether differences are independent of or related to size.

In table 2-2 we find clear differences in research cooperation related to company size. Seventy-seven percent of the smallest companies, 54 percent of the middle-sized, and 38 percent of the largest had no research cooperation. Of the smallest companies only 23 percent thus claimed to have had any significant research cooperation, compared with 62 percent of the largest companies. Systematic and organized cooperation characterizes 6 percent of the smallest, 13 percent of the middle-sized, and 31 percent of the largest, while spontaneous and temporary cooperation is the case for 17 percent of the smallest, 33 percent of the middle-sized, and 31 percent of the largest. Both the prevalence of and the degree of cooperation thus increase with an increase in company size, according to our data.

In our analysis of the importance of research cooperation for R&D success, we therefore carried out a separate analysis for each company size. This was made necessary by the results of the preceding analysis, which indicates that company size is associated with differences in both research cooperation and R&D success. Any attempt to study the relationship between cooperation and success must therefore allow for differences in company size as an influencing factor.

In table 2-3 the results for the smallest companies are given. As we have mentioned before, 77 percent of these companies stated that they had had no research cooperation, 17 percent that they had had spontaneous and temporary cooperation, and 6 percent systematic and organized cooperation. The majority

Table 2-2
Relationship between Company Size and Research Cooperation

Variable	Company Size (Total Number of Employees 1977)		
	Up to 9	*10-49*	*50+*
Number of companies	53 (38)	48 (34)	39 (28)
No research cooperation	41 (77)	26 (54)	15 (38)
Spontaneous and temporary cooperation	9 (17)	16 (33)	12 (31)
Systematic and organized cooperation	3 (6)	6 (13)	12 (31)

Note: Numbers in parentheses are percentages.

Table 2-3
Relationship between Research Cooperation and Success in R&D for Companies with Fewer than Nine Employees

	Research Cooperation		
Variable	No External	Spontaneous and Temporary	Systematic and Organized
Number of companies	41	9	3
(percent)	(77)	(17)	(6)
Total number of new products per year and company during the ten-year period	0.06	0.11	0.2
Total number of patented new products per year and company during the ten-year period	0.02	0.01	0.07
Average number of new products per year and company in relation to R&D effort (number of R&D employees)	0.27	0.32	0.19
Average number of patented new products per year and company in relation to R&D effort (number of R&D employees)	0.08	0.03	0.06

of the smallest companies thus evidently believed that they did not need or could not afford external help in their R&D and instead relied to a large extent on their own competence and resources.

Small companies that had spontaneous and temporary cooperation were more successful than those with systematic and organized cooperation, if we use our quantitative measure of R&D success, namely the total number of new products in relation to R&D effort (0.32 compared with 0.19). The corresponding figure for companies without external research cooperation is intermediate to these values (0.27).

If we instead use our qualitative measure of success—the number of patented products in relation to R&D effort—companies with systematic and organized research cooperation have been more successful than companies with spontaneous and temporary cooperation (0.06 compared with 0.03). They have, however, been less successful in this respect than companies without research cooperation (0.06 compared with 0.08).

A reasonable interpretation of these results is that a large number of the smallest companies during the period studied had a surplus of promising ideas of their own and sufficient R&D resources to develop these ideas internally.

For small companies interested in research cooperation, a good R&D strategy for the purpose of achieving a large number of new products appears to be spontaneous and temporary cooperation. This strategy of cooperation appears attractive from the point of view of gaining many new ideas from a variety of sources at low cost. From the point of view of finding products with a high level of technological innovation and great patentability, the alternative cooperation strategy of emphasizing systematic and organized cooperation appears to have been more successful.

The results for middle-sized companies are in table 2-4. The distribution of companies with regard to different types of research cooperation in this case is more uniform than in the case of the smallest companies. Fifty-four percent claimed to have had no external research cooperation, 33 percent to have had spontaneous and temporary cooperation, and 13 percent systematic and organized. As for the smallest companies, many had no external research cooperation, but the number of companies with either spontaneous and temporary or systematic and organized cooperation was greater.

We shall again begin our analysis by looking at companies without research cooperation. Companies with spontaneous and temporary cooperation in this size group have also been considerably more successful than companies with systematic and organized cooperation in achieving a large number of new products in relation to their R&D effort (0.20 compared with 0.12). Compared with companies without research cooperation they have also been more

Table 2-4
Relationship between Research Cooperation and Success in R&D for Companies with Ten to Forty-nine Employees

	Research Cooperation		
Variable	No External	Spontaneous and Temporary	Systematic and Organized
Number of companies (percent)	26 (54)	16 (33)	6 (13)
Total number of new products per year and company during the ten-year period	0.08	0.16	0.2
Total number of patented new products per year and company during the ten-year period	0.02	0.03	0.13
Average number of new products per year and company in relation to R&D effort (number of R&D employees)	0.13	0.20	0.12
Average number of patented new products per year and company in relation to R&D effort (number of R&D employees)	0.04	0.03	0.08

successful in this respect (0.20 compared with 0.13). With regard to the number of patented products in relation to R&D effort, we again find that companies with systematic and organized cooperation have been more successful than both companies with spontaneous and temporary cooperation and companies without cooperation (0.08 compared with 0.03 and 0.04).

We thus find that tendencies for the middle-sized companies are similar to those for the smallest companies, with regard to the relationships between research cooperation and R&D success. A majority of the companies (54 percent) have relied completely on their internal R&D resources, but the proportion is smaller than in the case of the smallest companies (77 percent).

We also find that it seems to have ben a good strategy for companies in search of many new products to emphasize spontaneous and temporary rather than systematic and organized cooperation. This latter strategy, however, as in the case of the smallest companies, is associated with a larger number of patented products in relation to R&D effort than either spontaneous and temporary cooperation or no cooperation.

The findings on the largest companies are given in table 2-5. The table indicates that a large number (39 percent) have not emphasized external research cooperation, but the proportion is lower than among the smallest (77 percent) and middle-sized companies (54 percent).

Table 2-5
Relationship between Research Cooperation and Success in R&D for Companies with More than Forty-nine Employees

	Research Cooperation		
Variable	*No External*	*Spontaneous and Temporary*	*Systematic and Organized*
Number of companies (percent)	15 (39)	11 (29)	12 (32)
Total number of new products per year and company during the ten-year period	0.09	0.16	0.29
Total number of patented new products per year and company during the ten-year period	0.04	0.06	0.2
Average number of new products per year and company in relation to R&D effort (number of R&D employees)	0.04	0.05	0.02
Average number of patented new products per year and company in relation to R&D effort (number of R&D employees)	0.02	0.02	0.01

As in the case of the smallest and middle-sized companies, large companies with spontaneous and temporary cooperation have achieved more new products in relation to R&D effort than companies with no external cooperation or with systematic and organized cooperation (0.05 compared with 0.04 and 0.02).

With regard to number of patented products in relation to R&D effort, systematic and organized cooperation for the largest companies is not associated with greater success than is spontaneous and temporary cooperation (0.01 compared with 0.02). Systematic and organized cooperation is also associated with less success with regard to patented products than no cooperation (0.02).

For the largest companies as for the smaller companies, spontaneous and temporary cooperation is associated with a larger number of new products in relation to R&D effort than is systematic and organized cooperation or no cooperation. Contrary to what we have found for the smaller companies, systematic and organized cooperation is not associated with a larger number of patented products in relation to R&D effort than for the other strategies.

Summary

This study deals with research and development strategies for 140 Swedish companies manufacturing farm machines. The work is part of a long-term research project focusing on company strategies for R&D in Sweden.

On the basis of previous studies (Nyström 1977a, b, 1978a, b), certain aspects of R&D strategies are emphasized. In particular, external research cooperation in our earlier work appeared to be strongly associated with success in R&D and was therefore stressed in this study. This cooperation refers to the R&D contacts that companies have outside the organization, for instance cooperation with universities, research institutes, consultants, or independent inventors.

Other studies and our own previous work indicate that size is an important determinant of success in R&D. Since differences in size may well also be associated with differences in research cooperation, there is additional reason for us to begin with size in analyzing company strategies for R&D. Apart from this, it is interesting to investigate whether our data are consistent with other attempts to correlate company size with R&D success.

Our data show that company size is consistently related to success in R&D. Success, then, is measured both by the total number of new products and the number of new products that are patented. A larger number of products in relation to R&D effort is used as a quantitative index of success, while a larger number of patented products in relation to effort is used as a qualitative index. Both the total number of products and the number of patented products in relation to our data decreases with increased company size. Size in turn is measured by total number of employees in the companies studied, and R&D

effort by the time devoted to research activities. Small companies, according to our analysis, have thus been consistently more successful than large companies in R&D.

When we look at the relationships between size and research cooperation, we find that larger companies consistently make greater use of research cooperation than smaller companies, which rely to a much greater extent on their internal research competence. When looking at the relationships between research cooperation and R&D success, we therefore need to consider differences in company size.

In our analysis of research cooperation we classified companies in three categories, based on total number of employees. Small companies had up to nine employees, middle-sized companies from ten to forty-nine, and large companies had fifty or more.

We also distinguished between three types of research cooperation. Companies were classified as having little or no external cooperation, spontaneous and temporary cooperation, or systematic and organized cooperation.

Our results indicate that for all sizes of companies, spontaneous and temporary cooperation was a good strategy for finding many new products. Systematic and organized cooperation, on the other hand, for all companies except the largest ones, was apparently a better strategy for finding products with a high level of technological innovation, as measured by patents. At the same time many companies, particularly the smaller ones, seem to have had enough own ideas and ability to develop these ideas into new products to be successful with little or no external R&D cooperation.

References

Galbraith, J.K. 1952. *American capitalism.* Boston: Houghton Mifflin.
Holmström, S. 1976. Förskjutning från manuellt arbete till maskinarbete i jordbruket. *Meddelande från Jordbrukets utredningsinstitut* 4.
Jewkes, J., Sawers, D., and Stillerman, R. 1958. *The sources of invention.* London: MacMillan.
Lantbruksstyrelsen. 1976. Förteckning över ett antal svenska firmor som tillverkar eller försäljer maskiner, redskap m m for jordbruket 1976. Mimeographed.
Lindström, C. 1972. Företagets storlek och belägenhet som determinanter för dess uppfinningsaktivitet. Umeå universitet, institutionen för företagsekonomi.
_____.1971. Stordriftsfördelar i forskning och utveckling. In D. Ramström, ed., *Mindre företag: problem och villkor.* Lund: Prisma, pp. 122-132.
Nilsson, B. 1978. Är jordbruket övermekaniserat? *Lantbruksnytt* nr 4.

Nyström, H. 1977a. *Company strategies for research and development.* Rapport från institutionen för ekonomi och statistik nr 107. Uppsala: Sveriges lantbruksuniversitet.

———. 1977b. Strategier för FoU. *Ekonomen* 7:7-10.

———. 1978a. Company strategies for research and development. In M. Baker, ed., *Industrial innovation.* London: MacMillan.

———. 1978b. *Creativity and innovation.* New York: Wiley.

Scherer, F.M. 1965. Firm size, market structure, opportunity and the output of patented inventions. *American Economic Review* 55:5.

3

R&D, Corporate Growth, and Diversification

Ove Granstrand

The empirical observations presented here have been made in connection with a larger study of corporate behavior with respect to R&D and technological innovation. This study includes several hundred interviews with people in R&D, marketing, and top management positions in eight corporations in Sweden. These corporations are large (with respect to Swedish conditions), diversified, and with one exception multinational. They represent different industries and technologies and the pseudo names for them used here reflect this. However, the corporations are more diversified and less bound to a specific industry than their pseudo names suggest.

The following abbreviations are used:

Cn-Corp	Chemical Corporation
El-Corp	Electronics Corporation
E&F-Corp	Engineering and Food Corporation
E&S-Corp	Engineering and Steel Corporation
M&M-Corp	Mining and Metallurgy Corporation
P&P-Corp	Pulp and Paper Corporation
Ph-Corp	Pharmaceutical Corporation
Tp-Corp	Transportation Corporation

El-Subs, Ph-Subs, and Tp-Subs (within aerospace) refer to specific subsidiaries in the corresponding corporations.

Other abbreviations used are:

S&T	Science and technology
R&D	Research and development
CMD	Corporate managing director
MNC	Multinational corporation

Aims and Limits

The primary aim of this chapter is to explore the relationship between, a corporation's R&D operations and its growth and diversification. A secondary aim is to present empirical observations to a larger extent than would have been needed for pursuing the lines of discussion. This has been done because there is a need for empirical studies of industrial R&D that go beyond just casual observations or supporting evidence for a line of argument, for example, in policymaking. A third aim, or rather a hope, is to bridge the gap between macro- and microeconomics. The study is not on the sector level but provides some insight into the differences of corporate behavior in different sectors of industry with respect to R&D and technological innovation.

The study is mainly qualitative although available data in some cases have made quantitative calculations possible. Concerning degrees of diversification, only rough first approximations have been made and further studies are required to develop more accurate measures of diversification and continuity in corporate and technological development.

Concepts

Growth in this context is taken to mean growth in total sales unless otherwise specified (for example, growth in number of employees or growth of R&D budgets). Obviously, total sales of a corporation may decrease for a period but growth is the dominating pattern. Growth is in some contexts taken to include other quantitative as well as qualitative aspects of corporate development.

The term *diversification* is used for the expansion of the variety of a corporation's products. Thus the sales of an old product on a new foreign market are treated not as a case of diversification but as a case of internationalization. Diversification is also used for R&D operations to denote expansion of competencies into different technological fields.

Diversification of a corporation is difficult to specify and measure. The concept pertains to similarities and dissimilarities among product characteristics and to the structure of sales of different products in the corporation. Concepts such as continuity and change, the newness of a product, and incremental versus radical innovation are involved, as well as concepts such as the importance of different products to a corporation. With respect to product characteristics, diversification has been specified on three levels: sector level, product area level, and product line level. These levels are sometimes diffuse as are most conceptualizations of technologies, products, and markets; but *sector* refers roughly to the highest level of a standard industry classification, *product line* to a parameterizable set of products, and *product area* to a set of product lines associated to one another in some dimension and lying within a sector. Innovations and diversifications leading into new product areas are referred to as semiradical.

Another way of specifying diversification is to relate it to the position of the corporation in the industrial input-output system and talk about horizontal integration, vertical integration, backward and forward and conglomerate diversification. It is also possible to qualitatively judge degrees of diversification by a product's relatedness to existing technologies and markets. Such a relatedness is multidimensional and has to be weighed if a single measure is sought. Here semirelated diversification refers to a new product area for a category of customers indirectly connected to the corporation in the input-output system. Finally the degree of diversification with respect to the structure of sales is roughly measured as the proportion of sales in product areas other than the largest to total sales, as defined by the product divisionalization of the corporation. This is but a rough indicator of quantitative extent of diversification. For instance, integration could have been separated from diversification and measured through the ratio of value added to total sales. However, most of what is said and discussed is not very sensitive to these quantifications.

Empirical Findings

*History of Corporate Growth
and Diversification*

The corporations studied show a wide variety of patterns of development both internally and among the corporations. A summary of the main features of this variation appears in appendix 3A.

In the initial phase of development a corporation is sensitive to many specific external conditions. It is not surprising, therefore, to find a variety of initial development patterns. Some corporations were slow starters and then picked up, while others grew fast and then experienced a stagnation; some had to wait several years for profits, while others were profitable from the start; some could internationalize at an early stage, while others had to diversify. A period of successful development could unexpectedly be followed by stagnation or severe decline, and so on. The initial sensitivity of a corporation to external conditions yields many small margins and rapid changes and thus challenges the versatility of top management. Together with random factors, the characteristics of a small group of leading actors in the corporations therefore provide quite a few explanations of initial corporate development.

Although the corporations develop within many dimensions, there are often discernible time periods or phases during which development within one dimension or in one particular respect dominates (or at least is later conceived of as having dominated). Such a period in growth process is often characterized by bottlenecks, although there are several other bases for division into periods. Thus the development in E&F-Corp may roughly be described as consisting of periods in which the sequence of production, financing, marketing, and R&D were dominating problem areas. This sequence results in a kind of sequential problem solving applied to different organizational functions. The periods of a particular orientation sometimes span more than a decade. This point may be further illustrated by the history of E&S-Corp as shown in table 3-1. Naturally the sequence of the main characteristics in different periods may change, but one often finds an initial sequence of problems in production, financing, and distribution.

All corporations studied experienced World War II, and all except M&M-Corp and Tp-Corp also experienced World War I. With the exception of one corporation these wars favored both growth and diversification; World War II especially favored long-term growth. Diversifications, due to shortages of supply and the production of defense material, were mostly temporary, although some technological substitutions were retained after the war as well as some permanent acquisitions made because of the war. The post-World War II demand in

Table 3-1
Periods in the History of E&S-Corp

Period	Characterization
1907-1918	Establishment, internationalization Production and material problems No new designs No severe financing problems Integration backward into steel Favored by World War I
1918 to mid twenties	Concentrated R&D effort Continued buildup of sales organization Depression
Mid twenties to early thirties	Development of sales organization Rationalization of production Acquisition of foreign bearing companies and a machine company 1928-1930
Thirties	Expansion of application Consolidation
1939-1945	Wartime problems Concentration on production
1945 to early sixties	Buildup and modernization of production facilities after World War II Post-World War II growth Acquisition of steel companies
Early sixties to early seventies	Emerging emphasis on coordination and diversification Acquisition of a large foreign bearing company and a tool company
Early seventies	Coordination and strengthening of production and R&D Continued diversification Acquisition of a tool company.

combination with the preparations for it during World War II was highly significant for the growth of most of the corporations. A growth in sales has been sustained ever since among the corporations, with inflation as a significant additional factor in the 1970s. The number of employees is, however, leveling off in several cases, thus yielding growth also with respect to sales per employee.

When thinking about changes in products and the product sales structure of corporations, one must consider several levels. On the level of aggregate product areas, the highest level of standard industry classifications, none of the corporations studied has completely changed its sector of industry; Ch-Corp is still in chemistry, E&F-Corp is still in engineering, and so on. Some corporations have at some time left or entered some other branch of industry, and some corporations have diversified permanently into some additional industry. P&P-Corp at some time during its three centuries left the iron industry and E&S-Corp entered the steel industry early; M&M-Corp recently entered and left the engineering industry, but it has also entered the chemical industry seemingly permanently. The continuity of the more specific product areas is still high. The original product areas remain; Ch-Corp still produces fertilizers, El-Corp light bulbs, E&F-Corp separators and milking machines, E&S-Corp ball bearings, M&M-Corp copper, P&P-Corp steel manufacture and wood products, Ph-Corp heart medicines, and Tp-Corp passenger cars. With respect to proportion of total sales, there have been shifts in the largest product area positions in Ch-Corp, El-Corp, P&P-Corp, and Tp-Corp, although the shift was temporary in Tp-Corp. These shifts of largest product area happen over long periods of time; at the same time the second largest product area may still be going strong. In Ph-Corp, with highly R&D-based products, the largest product is still growing after thirty years although it will probably be passed in the next few years.

Examples of temporary or nonremaining diversifications are sheet metal (E&F-Corp), jewelry and vermouth (E&S-Corp through compensatory business in the thirties), lead batteries (M&M-Corp), packaging machines (P&P-Corp, a new attempt at integration into packaging has been done through an acquisition in 1975), sweeteners (Ph-Corp during World War II), and ventilation equipment (Tp-Corp). Examples of remaining diversifications are plastics (Ch-Corp), radios (El-Corp), heat exchangers (E&F-Corp), fertilizers (M&M-Corp), bleaching chemicals (P&P-Corp), penicillin (Ph-Corp), and jet engines (Tp-Corp).

It is possible for diversification to become even more specific at the product line level, in which case technological and product substitutions and differentiations become more frequent and continuity decreases. Although all the original product areas of the corporations remain, the original product types and variants do not unless they have become standardized bulk products, such as standard chemicals, metals, and pulp. Separators, bearings, light bulbs, heart medicines, and cars have changed. Product ranges have grown in different technical parameters, components and materials have developed, performance has improved in different respects, new product generations have emerged. To describe patterns of technological and market changes on this level would lead too far astray, but a few observations may be made. First, product life cycles are commonly looked on as becoming shorter and shorter. This holds true in electronics, for example, but not in pharmaceuticals because of increased societal control which also affects chemicals used as fertilizers. Second, there are technical parameters such as size, effect, speed, reliability, and purity within which product technology has developed more or less continuously. Third, product and process changes involve to an increasing extent many different technologies. Technologies and scientific disciplines in turn differentiate and amalgamate, but their identities as coherent intellectual fields are connected to the university system and do not change as fast. Thus many corporations find themselves working in a conglomerate of technologies. Fourth, there are differences with respect to the characteristics of technological change in different product areas. The electronics field is characterized by rapid change intertwined with many other fields such as solid-state physics, chemistry, optics, and mathematics. Pharmaceutical research is often accidental, and the discovery of new pharmacological principles may lead to indications for several diseases. Within traditional process industries, technological change is not so rapid but significant long-run trends occur, for example, transitions from sulfite to sulfate pulp or transitions from basic to acid steel. (On a corporate level the changes may be drastic when the installation of new process equipment is required.)

Finally, the relationship between product changes and process changes differs both over time and among product areas. Disregarding for the moment the possible faults in the product-process distinction, depending on which step in a refinement chain one is considering, one can generally say of standard chemicals, bearings, and passenger cars that products have changed less (in some sense) than the way they are produced. Production technologies also change continually in the engineering industry with respect to parameters such as size, operation times, degree of automation, and reliability. It almost goes without saying that products subjected to different patterns of technological change are an integral part of most processes.

Diversification on any level of specification regarding sector of industry, product area, and product line may also be characterized according to changes in corresponding markets. (As is well known, one of the main motives behind diversification is to weaken dependence on changes in existing market conditions.) Some corporations or parts thereof, such as Ch-Corp, have diversified from producer to consumer markets, and others such as El-Corp from consumer to producer markets. Tp-Corp operated in both consumer and producer markets almost from the beginning (passenger cars and trucks) and has diversified into several other producer markets (car, marine, and aircraft engines, construction, agriculture and forestry machines) and into an additional consumer market (recreational items). The markets for chemicals do not have the same business cycles as the markets for metals, thereby weakening the dependence of M&M-Corp on the heavy fluctuations in the latter type of markets. Ph-Corp has diversified within pharmaceuticals but has also tried diversification outside this sector. Pharmaceuticals are in general not sensitive to business cycles but run a risk (varying over time) of becoming nationalized. E&S-Corp has diversified extensively within bearings and thereby acquired a certain insensitivity to market changes since different bearings are used in many industries. This also holds true for separators with E&F-Corp.

An aggregate picture of corporate diversifications is shown in table 3-2. It is difficult to make overall judgments in terms of successes or failures of these diversifications except in a few cases. Several remarks, however, are relevant in connection with the table. First, most corporations employ mixed strategies for diversification over time, but acquisition of companies was the dominating means of diversification in all corporations except El-Corp and Ph-Corp, which are also the most R&D intensive (see table 3-5). Acquisitions of companies have often come about as a result of planning in which diversification has sometimes been a side effect, since the acquired company has contained products that were not the primary target for the transaction. In some cases the acquisition came about as the result of an offer to acquire or a perceived threat that a competitor would make the acquisition. If there is a real choice between diversification through acquisitions or internal R&D, important factors include the availability of external technology and internal R&D resources, technological and competi-

Table 3-2
Type and Strategies for Product Diversification Still in Effect in 1975 in the Corporations Studied

Type of Diver-sification	Strategies for diversification			
	Acquisition of Companies	Acquisition of Technology	Internal R&D with External Cooperation	Pure Internal R&D
Integration backward	E&S, P&P, Tp	El		
Integration forward	M&M, P&P	P&P	El, E&F	El
Horizontal integration	E&F	Ch, E&F, Ph	E&F, Ph, Tp	E&S, M&M, Ph, Tp
Semirelated	Ch, El, E&F, E&S, M&M, Tp	Ch, E&F	El, E&F Ph, Tp	Ch, El, E&F
Weakly related	Tp		El	

Note: The cases of diversification in Ch-Corp and El-Corp are not completely covered due to the large degree of diversification in these corporations.

tive positions, synergies, and, not least, experiences and preferences within top management. Both ways are difficult, and explanations of successes and failures are not generally applicable. There are examples of overoptimism regarding possibilities of diversifying through both acquisitions of companies and internal R&D. Mixed strategies of some kind seem to have greater prospects of success. The strategies may be mixed in several modes. In a way acquisitions of patents, know-how, and R&D personnel or R&D cooperation with customers, competitors, suppliers, universities, inventors, and so on, constitute mixed strategies. Ch-Corp and E&F-Corp throughout their histories have ben talented acquirers of patents and know-how, and Ph-Corp has been talented in cooperating with universities and other companies. Another way to mix strategies is to acquire small, innovative companies in a proper stage of development, after they have passed the risky years of initial development but before they become stable. It may also be appropriate with a mixed strategy consisting of two strategies applied at the same time, for instance, when a company is acquired with R&D overlapping internal R&D. It is hard to achieve R&D synergies with an acquired company if R&D projects or R&D people are transferred; but if firm coordination is not applied at once, fruitful internal R&D competition may result.

The patterns of diversification differ greatly among the corporations in different sectors of industry. Diversification in the pharmaceutical field in Ph-Corp has proceeded along several lines. With the exception of the initiation of these lines, diversification has been rather coherent within pharmaceuticals although new competencies have been added and new areas of disease have been worked on. Raw material-based corporations such as M&M-Corp and P&P-Corp integrate mainly forward and backward and mainly through acquisitions of assets, companies, and technology. The raw material base has remained the same for M&M-Corp and P&P-Corp, although locations of mines and forests have changed and even become slightly internationalized, while the raw material base of Ch-Corp has changed completely. The securing and utilization of raw materials and by-products in production have been marked features of diversification in at least M&M-Corp and Ch-Corp. The composition of ores and other raw materials and available chemical processes, in combination with economy, have then largely determined the course of diversification. Diversification in Ch-Corp and El-Corp has been largely multidirectional. One way to explain this is to refer to the combinatorial nature of corresponding technologies, the wide applicability of products and knowledge, and the high R&D intensity. On the other hand, the products of E&F-Corp and E&S-Corp also had a wide applicability, with many opportunities for diversification outside the product area. However, these products were sold on oligopolistic producer markets, while El-Corp initially operated on consumer markets, E&S-Corp has had opportunities to integrate forward, but there has been internal resistance to it, and the qualified buyers present external resistance to integration forward. E&F-Corp, on the other hand, has integrated forward through becoming systems oriented.

This was stimulated by an increased systems orientation among some customers, and E&F-Corp in some early cases took the opportunity to supply the buyers with an extended range of components and know-how. Tp-Corp has diversified in a reverse way to systems orientation in an area in which it originally worked, the systems technology of design and assembly of cars, and then has integrated backward into R&D and production of central components. All the engineering corporations have also integrated horizontally, in the sense that the original product line has been supplemented with additional but closely related product lines for new markets, for example, ball and roller bearings, centrifugal separators and decanter centrifuges, passenger cars and trucks.

Vertical integration always poses the problem whether to deliver to competitors or compete with customers. This problem may restrain both marketing operations for intermediate products and integration forward. Thus, for example, E&S-Corp had a splendid opportunity to integrate forward into car production in the thirties with the aid of Tp-Corp, which was then a subsidiary of the E&S-Corp. For several reasons, one of which was to avoid competition with customers, this opportunity to diversify was not taken. Instead Tp-Corp was literally given away to its shareholders. (The Corporate Managing Director at E&S-Corp did not "believe in" Tp-Corp and was about to sell it to a U.S. automobile company but was persuaded not to by the Corporate Managing Director of Tp-Corp, who was a former sales manager at E&S-Corp.) Tp-Corp has in turn confronted the issue of integrating forward, for example into shipbuilding, but has refrained on similar grounds. Thus a kind of pull-push pattern occurs over time with respect to products and markets, and corporations choose different ways of deciding their limitations in this process.

History of Corporate R&D

El-Corp, E&F-Corp and E&S-Corp were clearly based on product inventions, all of which were product improvements rather than radically new products. Ch-Corp, and in a way P&P-Corp at the start of the seventeenth century were based on foreign technology and domestic raw materials. The discovery of ore deposits with a new ore-prospecting technique made the way for M&M-Corp. Tp-Corp was started on the basis of a domestic design, a comparative advantage in labor costs, and the pattern of foreign car industry. Ph-Corp was finally based on a permissive change in legislation.

In the economic combination of productive factors many circumstances were naturally influential, not least managerial talent. Technological change played important but different roles in all corporations except Ph-Corp. A question then is what roles technological change and, more specifically, corporate R&D operations played in the subsequent development of the corporations and vice versa, that is, how corporate development influenced R&D. A summary with respect to history of corporate R&D is given in appendix 3B. It is difficult to summarize, considering the many parts of a corporation that emerge and the many small ongoing improvements, for example, in production technologies.

In-house R&D has existed more or less from the start in all corporations except Ch-Corp, P&P-Corp, and Ph-Corp. The nature and intensity of R&D operations of course varied, and some work would hardly qualify for the label R&D by more rigorous standards. A kind of continuous "grass-roots R&D," often not especially recognized and organized in R&D departments, has played an important role in many corporations, however. Continuous technological development is not so easily recognized or assessed and is the result of small achievements by technologists in many positions in a corporation. The production function in M&M-Corp, Ch-Corp, and the engineering corporations and the marketing function in E&F-Corp and E&S-Corp have traditionally attracted many good technologists, and much R&D work has been done without being organized in an R&D department or the like.

To a large extent, earlier R&D work was a consequence of existing products, production facilities and raw materials. Organized R&D efforts for corporate growth and diversification are sometimes thought of as a new role for industrial R&D, having emerged after World War II. In the corporations studied, however, R&D efforts yielding significant diversifications apart from ongoing improvements were carried through before World War II in at least Ch-Corp, El-Corp and E&S-Corp. On the other hand, World War II was followed by significant changes in the way industrial R&D was organized and regarded. Most corporations extended their engagements in R&D during and after the war. At least six of the eight corporations built new laboratories and intensified R&D during the forties, four corporations made significant R&D advances during the forties, and as a result of the war R&D also grew up in foreign subsidiaries.

Although R&D in general has grown and diversified since World War II, there are variations among the corporations. Periods of intensified R&D have been followed by stagnation or decline and then by renewal. In Ch-Corp in the fifties much hope and many management commitments were connected with a large project. After a new Corporate Managing Director came, the project was stopped. R&D resources were transferred to other areas, and R&D temporarily declined for several years. The wave of divisionalizations in the late sixties implied in some cases a temporary decline in R&D. In the engineering corporations' periods of stagnation have occurred in the established product areas, in the sense that R&D has become increasingly oriented around some parameters for improvements. This phenomenon seems to depend more on the R&D management and staff than on the product or technology.

After World War II the status of R&D as an organizational function on a par with production and marketing has grown. R&D management and organization philosophies have varied over time. In the fifties and the early sixties R&D was often believed to be an almost sure generator of valuable results for corporate growth and diversification. These beliefs were later followed by pressures to control R&D more firmly and integrate it more closely with other functions, especially marketing. The preferences for centralized or decentralized R&D, large or small R&D units, internal or cooperative R&D, have changed. R&D management and organization philosophies also vary greatly among sectors of industry and among corporations.

An overall historical pattern is that corporate R&D has become larger, more diversified, and more internationalized as the corporation grows, diversifies, and internationalizes. On one hand, R&D is then a consequence of products and processes employed in the corporation; as soon as production is established there will be growing internal forces among employed technologists to perform R&D. Since the corporation is large, these forces are difficult for corporate management to eliminate. On the other hand, different features of corporate development are to a varying extent a consequence of R&D. Some cases of R&D-based diversifications are described in more detail in appendix 3C.

There have been shifts in or significant additions to the dominant technology of almost all corporations. For instance a progression of generation shifts from carbide engineers to polymer technologists in Ch-Corp; a progression of generation shifts in electrical engineering from vacuum tubes to transistors to integrated circuits to microcomputers in El-Corp; chemistry, biology, electronics, and systems engineering have been integrated in mechanical engineering in E&F-Corp; material scientists have been upgraded in E&S-Corp; metallurgists and chemists have been added to "the mining people" in M&M-Corp; biologists and mechanical engineers have been upgraded in P&P-Corp; a transition from chemistry to biology has taken place in Ph-Corp; and mechanical engineers have been supplemented in Tp-Corp. These changes in the portfolio of technological competences depend on external technological development and internal conditions, as, for instance, the rise of advocates or resistance among management and technologists.

Four kinds of diversification of R&D operations may be discerned. First, there is a diversification of competencies pertaining to the core technologies of a corporation, for instance the differentiation of polymer technology or tribology. This is "ordinary" specialization within a technology of decisive importance to the corporation. Second, there is a diversification pertaining to adjacent technologies. These adjacent technologies may concern the supporting technologies such as automation technology in production, surface chemistry for lubrication in a part of a product or materials technology. Corporate R&D often diversifies into adjacent technologies through an initial stage of perception of product problems followed by attempts to solve them by extending internal knowledge, often amateurishly, or hiring external R&D services. Third, there is substitution among different technologies, such as the transition from chemistry to biology in pharmaceutical research. Fourth, a new technology is "picked up" for the assessment of its potential benefit to the corporation or directly for creating new businesses; for example, Ch-Corp acquired polymer technology and Ph-Corp went into antibiotics.

The diversification into a new technology for new kinds of businesses is quite often evolutionary, with a progression over adjacent or substituting technologies. For instance, the need to preserve milk has led E&F-Corp into heating and cooling, in turn leading to heat exchangers, microwaves, the preservation of other types of food, and finally to a new packaging technology. (At present E&F-Corp has decided not to go into packaging.) The concept of evolutionary chains are too simplified though; rather, technologies advance along some lines, may then rest until combined with some other technologies, and then may advance a bit further.

Any typology of the diversification of R&D is vague, since conceptions of a technology are diffuse and changing. Confluences and combinations occur. Strictly speaking, diversification of R&D should be considered to decrease if a combination of two technologies gains coherence and recognition. Many corporations have encountered different environmental problems in the recent decade and have developed countermeasures in the form of corrective technologies. New competences had to be acquired, and perceptions of which technologies were adjacent and relevant changed rapidly. Thus the kind of diversification of R&D bred by environmentalism is hard to classify. It may not be considered a diversification at all since environmental technology has become recognized.

Growth and diversification of corporate R&D are strongly connected to external developments of technologies and markets. Common features, at least after World War II, are that R&D costs and R&D times for a new product are increasing (table 3-3), and the commercial product life is shortened. Although there are exceptions and the variations are large among different product areas, these general features are not due primarily to corporate development, nor are they inherent in the economy of corporate R&D operations.

Present Situation

Because of confidentiality and space limits it is not possible to be very specific about the present situation with regard to R&D, growth and diversification, and corresponding strategies (policies). A summary is shown in table 3-4. Aggregate corporate strategies (policies) all emphasize profitability and growth with some variation in relative strength. Internationalization is emphasized in most corporations, especially in Ch-Corp, M&M-Corp, and Ph-Corp. Diversification has been deemphasized in Ch-Corp, Ph-Corp, and Tp-Corp, while E&S-Corp, P&P-Corp and M&M-Corp place great emphasis on increased diversification which has to be synergetic. Finally, R&D is emphasized in all corporations, although less so in P&P-Corp. R&D began to stagnate briefly in the early seventies in Ch-Corp, E&F-Corp, and Tp-Corp. Research remains mostly domestic while development

Table 3-3
Examples of Statistics on Product R&D

Product	Corporation	Calendar Time	R&D Time[a] (years)	R&D Cost[a] (Sw.Cr. in millions)
Roller bearings	E&S-Corp	1918-1921	3.5	0.75
Local anaestheticum	Ph-Corp	1943-1948	4	b
New generation of passenger cars	Tp-Corp	1953-1956	3	50[c]
Hydraulic machines (pump, motor)	Tp-Subs	1962-1967	5	<10[d]
Jet engine	Tp-Subs	1962-1968	6	e
First generations of beta-blockers	Ph-Subs	1960-1967	7	7-8
Second generation of beta-blockers	Ph-Subs	1966-1975	9	30

[a]Beginnings and ends of R&D are diffuse. The figures apply to the period of a recognized industrial R&D project until serial production starts (for pharmaceuticals until marketable product). Sources are internal interviews and documents.

[b]The decisive discovery was made in 1943 by university researchers after several years of work. Sales still growing (1978).

[c]The development of a new generation of passenger cars for the future is estimated by the Tp-Corp to cost around 1 billion Sw.Cr. just using known technologies.

[d]Sales amounted to 48 million in 1977, 83 percent of which was for foreign markets. Still not break-even.

[e]Four million man-hours were required. The development work was, however, based on licensed technology. The figures apply to contract R&D for the Swedish Air Board. The complete development of a new jet engine generation for the future is estimated by the Tp-Subs to cost 3 billion Sw.Cr.

is increasingly internationalized in some corporations. There is also a trend toward concentrating R&D on certain product areas, although decisions on this subject differ among the corporations.

The reasons expressed for sustained growth are rather predictable, including references to rising labor costs, technological change, rising costs for R&D and marketing, growth as a necessary condition for enduring profitability, and the vicious circles related to stagnating growth. The reasons expressed for diversification are more varied and more product oriented but still rather conventional, including references to inadequate growth and profitability in existing product areas, the need to distribute political and economic risks, increased insensitivity to business cycles, utilization of excess resources, securing input or output markets or other advantages of vertical integration, achieving synergies in R&D, production, or marketing, filling gaps in or supplementing existing knowledge or

Table 3-4
Aggregate Figures of Present Situations of the Corporations Studied

Variable[a]	Corporation								Correlation[b] with R&D/Sales
	Ch	El	E&F	E&S	M&M	P&P	Ph	Tp	
1 Size of sales 1975 (rank)	6	1	4	3	5	8	7	2	0.42
2 Sales 1975/sales 1970	2.29	1.80	2.31	1.44	1.85	1.72	2.01	2.57	0.18
3 Sales/employee 1975[c]	273	116	197	112	229	211	199	217	−0.28
4 (Sales/employee 1975)/ (Sales/employee 1970)	2.05	1.63	1.95	1.58	1.32	1.73	1.64	1.58	−0.11
5 Cumulative profit/cumulative sales 1971-1975[d] (percent)	8.9	8.3	7.5	10.1	9.5	11.0	7.9	7.3	−0.65
6 Diversification 1975[e]	0.79	0.83	0.53	0.21	0.64	0.74	0.26	0.45	−0.28
7 Internationalization 1975[f]	0.18	0.77	0.63	0.79	0.07	0	0.51	0.27	0.48
8 R&D/sales 1975 (rank)	6	2	4	5	7	8	1	3	1.0

Note: Results are influenced by the marked profitable growth in raw material-based industries around 1974. Growth increased from 1973-1974 in Ch-Corp, M&M-Corp, and P&P-Corp—44 percent, 47 percent, and 52 percent respectively.
[a]Because of confidentiality only ranks are shown here for variables 1 and 8.
[b]All correlations are calculated on metric data.
[c]Thousands of Sw.Cr.
[d]Trading profit after depreciation but before taxes.
[e]$1 - (S/T)$ where S is sales in largest product area and T is total sales.
[f]Number of employees abroad/total number of employees.

product lines, engaging in promising fields, and finding profitable opportunities. When companies make acquisitions, an unwanted diversification may sometimes occur as a result. There has also been restructuring of industry, mainly above the corporate level, which has resulted in diversification. A majority of the corporations do not emphasize diversification, at least not on the level of new sectors or product areas. There is a widespread orientation toward the continued existence of traditional products such as steel, bearings, separators, forestry products, and passenger cars and toward the fact that these should constitute a backbone in corporate futures as well. Some of the more spectacular technological threats, which were conceived of previously, have not materialized as substitutions as rapidly as has been expected. Besides, many examples of diversification failures have emerged. Synergies, profit opportunities, and the like, have been overestimated, and the consequences of lack of knowledge in a new area and division of management attention have been underestimated. In the cases where diversification is sought, a basic effort is to preserve the natural unity of the corporation, and diversification must be related to the present state at least to a certain extent. There is then a strong orientation toward existing knowledge associated with existing productive resources. Although there have been periods of perceived high risks for radical material substitutions (plastics for wood), it is almost inconceivable for corporations such as P&P-Corp and M&M-Corp to sell raw material assets and buy their way into some area of quite different materials with quite different knowledge needs.

Many corporations want to view themselves as selling market-oriented systems of some kind, but the orientation toward existing technologies and components is strong. Ph-Corp, for instance, sometimes claims that it sells systems for medical treatment in different disease areas. Although there is an increased emphasis on providing physicians and hospitals with know-how, the pharmaceutical product and its associated information ledge is the basic component. Other technologies such as electronics, surgery, and artificial organs are not actual targets for diversification. They may be relevant but are considered too distant from present technologies or too exotic. Similar statements may be made about the other corporations. Admittedly, some are modest in their claims about systems orientation, but the implications for diversification of a systems-oriented strategy or business idea are, as a rule, inadequately explored and judged.

The present situation of growth and diversification of R&D differs greatly among corporations and among their product areas. To a large extent it is a consequence of corporate profitability, growth, diversification, and internationalization. Historically there are several instances of growing R&D due to good years and declining R&D due to bad years. An increased readiness to let R&D be less affected by business declines may be noticed in some corporations. These are corporations, on the other hand, that do not face severe business declines. For several reasons it is tempting to manipulate R&D resources temporarily, and it is a large step from a general readiness to invest in R&D to making annual budget decisions in times of increased scarcity of resources. Corporate growth is not infrequently a prerequisite to changing relative proportions in resource allocations. It is also possible to transfer R&D people to other functions.

Budgeting obviously determines growth and diversification of corporate R&D in a narrow sense. This is not the place to review R&D budgeting practices, which always involve political processes to some extent. In some cases R&D budgets are built up project for project, commonly when a small number of large projects are involved. Comparisons with the corporate past, competitors, and sector averages, often based on R&D as a percentage of sales, are influential factors. Thus there is a direct coupling of R&D growth to corporate growth, which is strong in some corporations, especially Ph-Corp.

Corporate strategy naturally affects growth and diversification of R&D. Ch-Corp, for example, explicitly emphasizes advanced knowledge, knowledge—intensive projects, and increased R&D efforts as means of increasing insensitivity to business cycles, which are thought to affect knowledge-intensive products to a lesser degree. M&M-Corp and P&P-Corp have emphasized R&D for better utilization of raw materials. R&D is a means for internationalization of Ph-Corp and for diversification of the E&S-Corp.

R&D is to a varying extent being given some new roles in corporate development. This does not necessarily imply a growth in R&D, but sometimes it does occur. Internal R&D, recruitment, cooperative R&D, acquisitions of patents and licenses, acquisitions of companies, and purchase of hardware with embodied technology have been the ways in which the corporations have in varying proportions been supplied with technological knowledge. Factors such as increased international competition and advancement of science and technology in general might make it more difficult to secure such a supply without internal R&D. Internal R&D may then be a means of attracting R&D people and partners for cooperative R&D as well as giving the corporation access to the science and technology community and a capability of utilizing technological information. The degree of external orientation and cooperation in the performance of R&D in most corporations is increasing for various reasons, including perceived increases of costs and risks in R&D and increases in the range of relevant technologies and in specialization. Increased external orientation is especially emphasized at the corporate level in Ch-Corp, M&M-Corp, and Ph-Corp.

Other factors that stimulate growth of internal R&D are dependence on labor, which stimulates changes to more technology-intensive production; more advanced customer technologies; societal pressures for safer products and processes; and the use of R&D as a means of competition. E&F-Corp dominates some of its markets and has for a long time skillfully acquired patents. These conditions might decrease the need to be technologically offensive, but the E&F-Corp still experiences intermittent technological leads and lags in relation to competitors; thus corresponding temporary demands on R&D arise. It is also

R&D, Corporate Growth, and Diversification

sometimes said that a corporation has to develop technology rather incrementally with a few radical advances now and then. A common pattern in electronics and engineering is to develop new generations of products with a series of advances and to make product improvements in between, often on the component level. Again a temporary growth of corporate R&D might result. Technologically advanced products do, however, require resources, which puts pressures on prices in order to keep profits up. In order to keep high prices, a high level of technology is often needed, at least in competitive industrial markets. Thus, in addition to budgeting, there is another element of mutual reinforcement between corporate growth and the growth of R&D.

The last category of indicators for present growth of corporate R&D has to do with factors inherent in technological change and R&D, which are in that sense self-reinforcing factors. (Research breeds research.) Roughly these concern the nature of R&D people on the one hand and the nature of R&D and technological change on the other. The ambitions of R&D people usually favor R&D growth. Such ambitions may also favor diversification of R&D because R&D people tend to disperse their resources among different problems and ideas in order to weaken competition and get recognition. On the other hand, management may favor science and technology areas, which are already established in the corporation.

The question to what extent there are self-reinforcing factors in the nature of R&D and technological change is a complicated one. There are many interesting processes of growth and diversification—in different fields of science and technology in general, in the involvement of corporate R&D in certain fields, in the course of a project, and among R&D teams and individuals. The conceptualizations of different fields are also fluid.

The stock of discovered researchable problems tends to increase with increased knowledge about the possibilities of combining knowledge. The necessary scale of an industrial R&D project may grow merely by combining known technologies. The development of a new generation of cars in Tp-Corp cost 50 million Sw.Cr. in the early fifties; today it might cost 1 billion Sw.Cr. with known technologies. (See table 3-3.)

There is an aging process in individual researchers as well as in research teams. A research field may also age in the sense that there are diminishing returns on additional research. Some of the technologies of the engineering corporations are referred to as mature technologies, meaning that R&D no longer pays off so well, at least not product R&D. On the other hand, it is hard to find a technology that is nearly perfect and without conceivable breakthroughs. It may very well be that R&D people in a field have stagnated or have become emotionally attached to certain solutions and that breakthroughs in the field are more likely to emanate from R&D in another field. But there is certainly an element of self-fulfillment in saying that a technology is mature and little will happen so that R&D may be kept low in the corresponding product area. Some corporations are influenced by theories about the product life cycle and the maturing of products and markets. The use of a milking strategy with little R&D for a product with low market growth and high market share clearly lowers the possibility for changes in the product technology.

Closely related to the question of self-reinforcing factors in the growth of R&D and technological change is the question of the effects of size. A distinction must be made here between the cumulative size of R&D efforts in a certain field and the size of R&D teams, departments, laboratories, and total R&D in a corporation or sector of industry. There is much to be said on this subject, and it is a common topic of discussion among R&D managers and others. There are widespread notions about critical or threshold sizes of an R&D effort, barriers to growth and diversification of R&D organizations, optimal sizes of an R&D team or an R&D laboratory, and so on. The actual situation, however, gives little support for estimations and generalizations of these notions. The research laboratories at the corporate level in El-Corp in the early seventies employed 2,250, 580, 440, 400, 330, 175, and 45 persons respectively. Although the work of these laboratories is diversified, they are to some extent involved in similar science and technology fields. The variations in R&D between different corporations in largely the same sector of industry are also large, at least in some sectors. In the forestry, pulp, and paper industry there is roughly one group with internal R&D, one group with largely collective R&D, and one group that performs almost no R&D. The steel unit of E&S-Corp spends little on R&D in comparison with sector averages. The explanations given for this are a relatively concentrated product program and early R&D success, but also the fact that the vertical integration has led to some stagnation after the initial upheaval. The variations in R&D averages in different sectors of industry are well known. References to the differing nature of different technologies are common, but many other factors accrue and this problem constitutes a research problem in itself. One illustration is that the size distribution of the cooperative research institutes in different sectors ranged in 1976 from around half a dozen people to around three hundred.

Size is hard to assess because many external employees may be involved temporarily. An extreme case is the R&D organization behind the combat aircraft Viggen; at one point in the sixties it consumed about 10 percent of all Swedish R&D funds and as a total R&D effort was comparable in relative size to the Apollo project in the United States. The large Viggen project was organized into a central coordination department with a series of contractors, among else Tp-Subs. What, then, is the significant measure of size of an R&D organization? This question also raises the question whether R&D management may be performed without performing in-house R&D, which some R&D managers argue is not possible in the long run. A parallel notion is that industrial R&D has to be located close to production and other business operations, although a certain degree of detachment is sometimes sought. (The corporate R&D laboratory of

E&S-Corp has a completely separate location in a country in which E&S-Corp has only minimal operations.)

Many factors account for these size variations. One is the variation in views on size effects. It is not uncommon for an R&D manager to view his department as being close to a critical size while corporate management views it as having a suitable size. The standards of evaluation may differ, and it is also possible but unlikely that both are true, using the same standard. The critical size may also depend on technological change. In electronics the relationships with material science, physics, chemistry, optics, and medicine are becoming closer, raising the critical size of a research laboratory in El-Corp. Involvement in systems technology is similarly said to require large and growing R&D efforts. It is possible to use subcontractors, a traditional strategy of Tp-Corp, but the growing complexity of preproduction favors internal R&D.

There is a certain consensus on the existence of optimal size of an R&D laboratory in a corporation. If a distribution of laboratory sizes gives no indication of such an optimum, it is possible that as an R&D laboratory grows an optimum will be passed, unless the optimum size is very large or is growing faster than the laboratory size does. However, in no case are there indications of such a passage. The R&D manager in Ph-Subs planned in 1964 to reach an optimal R&D unit of 150 people in 1975. Since then diversification into new and unexpected areas has occurred, and the size in 1975 is 175 employees and growth to 250 employees is planned. He estimates the critical size of an R&D unit in pharmaceutical research in the mid seventies at roughly 100-150 people, and the optimal size may be in the range 150-300 as estimated by R&D and corporate management. Problems are anticipated, since there are internal forces that must grow and diversify in the different R&D units of the corporation. A possible course of action is to concentrate within each R&D unit, increasing risk taking on subsidiary level, and possibly establish new units, which increases demands for coordination on corporate R&D management. Besides there is a fluctuating pattern in R&D costs when R&D projects reach the final expensive stages, and there is a fluctuating pattern in R&D output due to the aging of R&D teams and R&D fields. Thus there is R&D growth fluctuation on several levels.

There is actually not very much to say about limits on R&D size on the basis of observations. Sometimes there is a policy of undermanning and compressing time frames in R&D, but even in the absence of such policies there is always an inexhaustible stock of problems, which never constitute a limit to size. Another conception of limit to size is based on diminishing returns. Certainly additivity of R&D resources is low, and possibilities to make substitutions are limited. Two mediocre researchers do not equal one top researcher. Adding people to an R&D team may even be counterproductive. These are all intuitive notions used in specific situations, but they seldom give a basis for estimating any limits to size.

There is a consensus that drastic changes upward or downward in the size of a R&D unit are detrimental. Tp-Corp and E&F-Corp reorganized R&D in connection with diversification. Central R&D units were reduced, and the decline was reinforced by a lowered morale and loss of employees. Ph-Subs, on the other hand, is experiencing some limits to growth. Bioanalysis is to be expanded, pharmacology is to be doubled within an unspecified time, and medicine is to be doubled in five years. The rate of expansion, however, is determined by recruitment possibilities and the speed with which newcomers can be assimilated into the organization without damaging it.

Limits to changes in size obviously limit the use of temporary R&D efforts, like the one in E&S-Corp after World War I. Many factors are influential: economic, technological, organizational, social, physchological. Recruitment and transfer possibilities are important and differ among different fields. Mobility of R&D people also changes with age. A most effective limit to growth, finally, is the limits on individual learning.

More pluralistic and temporary forms of organizing and managing R&D are being employed. There is an increasing use of satellite organizations with semiexternal employees as inventors, consultants, and university researchers; increased degree of external orientation in general with cooperation, licensing, and know-how exchanges; increased use of semiautonomous innovation companies; and increased use of project organizations and R&D units on corporate, divisional, and regional levels. Phrases such as task forces, venture teams, and ad hoc groups as well as the basic idea behind a project organization indicate that temporary organizations are being used. Thus the organizational ways of conducting industrial R&D have also diversified.

Discussion

From the empirical observations a large and diversified set of topics emerge for discussion and relating to literature, and just a few topics may be touched on here.

At the corporate level the different patterns of corporate growth and development cannot be aggregated into a general progression of stages without significant simplification. Scott (1971) distinguishes between three stages of corporate development, as shown in table 3-5. For the corporations studied here this is but a rough model. The invention-based corporations initially developed around a single product or product line and then rapidly internationalized, while corporations based on raw materials or foreign technology initially diversified rapidly into at least two product lines. Diversified raw material sources, by-products in production, costs of transportation, and competition abroad favored early diversification, while a global advance in a product technology and a small domestic market favored early internationalization. Thus the origin of the corporation is important to consider as well as the size of the domestic

Table 3-5
Parts of Scott's Model of Stages of Corporate Development

Company Characteristics	Stage		
	Stage I	*Stage II*	*Stage III*
Product line	Single product or single line	Single product line	Multiple product lines
R&D	Not institutionalized, oriented by owner-manager	Increasingly institutionalized search for product or process improvements	Institutionalized search for new products as well as for improvements
Strategic choices	Needs of owner versus needs of firm	Degree of integration, market share objective, breadth of product line	Entry and exit from industries, allocation of resources by industry, rate of growth

Source: Derek F. Channon. *The Strategy and Structure of British Enterprise.* New York: Macmillan, 1973, p. 9.

market. Scott's model and many other models for corporate development are based on studies of U.S. corporations that have had uniquely large domestic markets. This fact limits the possibility of generalizing with the help of these models.

Relationships between R&D and Corporate Growth

The many means for and factors behind corporate development make it difficult to assess the impact of and on a certain factor such as R&D. When long periods of time and aggregate patterns are involved, causality is hard to identify. There is widespread belief that R&D fosters profitable growth. When resources are created by profitable growth caused by several factors, it is also likely that R&D is fostered thereby. This proposition is supported here in two ways. First, one may see how periods of significant change in corporate R&D budgets are preceded by periods of change in profitability and growth. R&D budgets or budget increases tend to be cut when profits decline even if the corporation continues to grow. Similarly, many investments in R&D facilities and projects have been made during good years. (Diversification efforts in both E&S-Corp and Tp-Corp were increased during some good years in the sixties.) Thus business cycles or fluctuations in general have a marked impact on R&D. Expectations may, however, displace the pattern somewhat and give rise to an increase in R&D during periods of emerging rise or decline in business.

Second, one may correlate time series of R&D and sales and estimate leads and lags. However, very little quantitative data are available on R&D, since corporate R&D statistics are of recent date and often confidential.

Correlations between the time series for sales $[S_t]$ and the time series for R&D displaced λ years $[R_{t+\lambda}]$ for E&F-Corp and Ph-Corp are shown in figure 3-1. The calculations are made on raw data, unadjusted for trends, inflation, and the like, only in order to indicate regions of high correlations, for example for $\lambda = 0$ and $\lambda = 1$. Calculations are not made for fewer than five observations, which does not allow one to see other peak regions for E&F-Corp due to lack of data. For the Ph-Corp correlations are also high for $\lambda = \pm 8$. Although the correlations for Ph-Corp generally are high, the correlograms suggest both a contemporary causal coupling and a lagged one between R&D and sales. The fact that sales lag behind R&D is in agreement with common belief. The extent of the time lag depends on the kind of R&D undertaken but also on how fast a new product is marketed and adopted and how fast sales increase. The lag for Ph-Corp is estimated, on the basis of the correlogram, to be seven to eight years, which may be compared with an R&D period of seven to nine years for pharmaceutical R&D in the 1960s, as shown in table 3-3.

Assuming a linear stochastic model for (1) the process of R&D-based growth of sales (2) the R&D budgeting process gives

$$S_{t+1} = \sum_{i \geq 0} r_{t-i} R_{t-i} + \sum_{i \geq 0} u_{t-i}, \qquad (3.1)$$

$$R_{t+1} = s_t S_t + v_t. \qquad (3.2)$$

Here u_t, v_t are lump variables with stochastic elements, and the summation is appropriately extended for nonnegative i. Under current practices of budgeting R&D, it is reasonable to assume that R&D is budgeted just one year ahead and partly based on present sales as reflected by $s_t \cdot s_t$ may fluctuate over time and even in the case of Ph-Corp; s_t is not a constant plus noise but has slowly drifted from more than 8 percent to more than 10 percent over the last fifteen years.

Equation 3.2 may be modified by including estimations of next year's sales, just as equation 3.1 may be further simplified by the rough assumption that all $r_{t-i} = 0$ except some $r_{t-\lambda}$, meaning that R&D in year $t-\lambda$ generates sales only in year $t+1$. The point here, however, is just to show how the two processes reinforce each other.

Substitution gives

$$S_{t+1} = \sum_{i \geq 0} r_{t-i} \cdot s_{t-i-1} \cdot S_{t-i-1} + \delta,$$

$$R_{t+1} = \sum_{i \geq 0} r_{t-i-1} \cdot s_t \cdot R_{t-i-1} + \epsilon. \qquad (3.3)$$

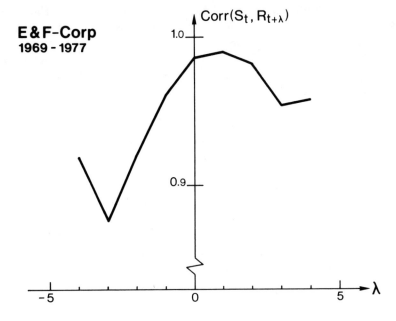

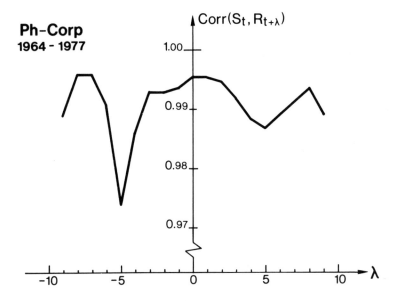

Figure 3-1. Correlations between Displaced Time Series for Sales and R&D

Here δ and ϵ are new lump variables independent of $[S_z]$ and $[R_t]$. Thus sales and R&D are self-reinforcing through the budgeting process, and if s_t is more or less time invariant and r_t changes slowly, the rate of reinforcement is roughly the same. (However, this has not been tested.) Adding to this the assumption that R&D in year $t-\lambda$ generates sales only in year $t+1$ implies that reinforcement is compounded each $\lambda+2$ years at approximately the same rate. Assumptions that a pulse in R&D yields a delayed pulse in sales, for example, may explain the slight peak in the correlogram of the Ph-Corp for $\lambda = 8$. (An underlying assumption is that other necessary investments are done as well. It is not uncommon, for example, for investments in marketing to be roughly of the same size as investments in R&D.)

The statement that growth fosters R&D is supported by Schmookler (1966) at the macro level. Schmookler also shows how inventive activities tend to lag behind investments in capital goods. This lag is also supported by this study, if only on the strength of a few examples. In the steel unit of the E&S-Corp investments are made with the next fifteen to twenty years in mind, while developing a process to marketable license takes four to eight years and developing a product may take less than five years. Materialized investment decisions thus largely determine subsequent R&D. In P&P-Corp a heavy investment was made in the early sixties in facilities for cardboard production, implying an integration forward. As a result of this integration a new R&D department for product development was started in the late sixties. A decision to acquire a supplier's product technology through capital investments, rather than through internal R&D, increases sales in the supplying sector of industry, which through the budgeting decisions in that industry causes corresponding product R&D to grow and lag behind capital investments at the aggregate level.

Schmookler further argues in favor of a demand-oriented rather than a supply-oriented theory behind growth of technology. On a micro level, this seems to be a justified view in early stages of corporate diversification through R&D; in later stages the situation becomes more supply oriented, at least for a period. Often initial periods in which technology is developed are followed by periods in which the areas of application for developed products and knowledge are extended. For instance, both E&F-Corp and E&S-Corp in the interwar years widened their areas of application while much product development work had been done before. The separators of E&F-Corp were universal process elements, and the rolling bearings of E&S-Corp were universal machine elements. Thus these corporations came into contact at an early stage with a wide variety of applications and customer problems, which stimulated development of a wide range of product variants in their specific product area. Sometimes the range of application of a new product is underestimated, as was the case with the original invention in E&S-Corp. At other times an entirely different application is accidentally found, as was the case when a pharmaceutical product for rhythm disturbances in the heart was found to have effects on high blood pressure.

Furthermore, an initial demand may lead to the acquisition or development of a product that is later marketed to other customers as well, as was the case with E&S-Corp, which integrated backward to steel and now markets steel products to external customers who may be competitors. In the Ph-Subs R&D was initially oriented toward large need areas within medical care, but as internal competences were built up, R&D became more oriented toward the existing competences. The systems orientation in E&F-Corp was fostered by demand conditions in the dairy industry in the early sixties, but the knowledge of how to handle combinations of machines has since been applied and adapted to other sectors of industry. Thus a pattern of "first pull, then push" occurs over time with respect to products and markets of a corporation.

The Relationship between R&D and Corporate Diversification

R&D has resulted in several marked diversifications. If the conditions pertaining to these diversifications are examined, one finds that in several cases of semiradical innovations leading to diversifications, there has been some direct external influence, for example, an idea from an external inventor (hydraulic machines, Tp-Subs) or ideas and advice from external researchers (several product lines in Ph-Corp). In other cases primary sources of influence have been internal, for example, from top management and R&D people (roller bearings, E&S-Corp; trucks, Tp-Corp) or just from R&D people (heat exchangers, E&F-Corp). Naturally, external market and technological conditions are influential in all cases, but the positions of persons who make interpretations and exercise direct influence are considered here. With respect to different types of diversification there is no general pattern, either. For example, in E&F-Corp the ideas of R&D people in combination with policy decisions have played the main role for diversification outside traditional areas and market factors for diversification within these areas. In E&S-Corp, on the other hand, there has been a different situation in this respect; external market and technological conditions have stimulated the development of new lines of bearings, while internal factors such as those concerning production technology have played a larger role in development within existing product lines.

Findings of this kind suggest that semiradical innovations and diversifications based on R&D are influenced by the existence of internal and external sources of ideas and impulses for diversification, a kind of permeability in the organization, and resource structure. The permeability in the organization pertains both to the susceptibility of an organization to external ideas and impulses and to the elasticity of an organization in the event that internal ideas and impulses lead to areas outside present operation (two-way permeability). This permeability, in turn, is influenced by policy decisions, top management attitudes, not-invented-here-effects among R&D people, and delineation of organizational rules and responsibilities. Such policy decisions may be to stay within certain business areas (not to go into the packaging business, E&F-Corp), to be restrictive with spin-offs (the Ph-Subs), to limit production (M&M-Corp), to develop adjacent technologies but stay within the original one (Ch-Corp), to diversify synergetically into present technologies (E&S-Corp) and to strive toward further refinement (P&P-Corp).

The permeability may differ with respect to internal or external ideas and impulses. A policy decision to concentrate on certain areas or to stay out of certain areas may limit internal initiatives, while external sources of influences may still be active. Bureaucratic procedures for evaluation may kill off external ideas, while improvements of existing products may take place internally without very much notice. There are several examples of how evolutionary steps starting from present operations lead to new products and markets (see also Salveson 1959). This process encounters many internal barriers that decrease the permeability and tend to confine the evolutionary process.

Summary

This chapter describes the history and current position of corporate growth, diversification, and corporate R&D in eight corporations in Sweden which represent different sectors of industry. The general conclusion is that a mutual interplay and a give-and-take relationship exist between corporate development and R&D.

There was a great degree of continuity in corporate development at both the sector and the product area levels with regard to diversification. The corporations used different mixed strategies for diversification, and internal R&D with external cooperation proved successful in several cases. R&D for diversification into new product areas proved not to be a strictly post-World War II phenomenon.

On the R&D level there has been an important kind of grass-roots R&D, whose results are not as easily recognized and assessed as indisputable innovations. Four kinds of diversification of R&D are discerned: differentiation of core technology, expansion into adjacent technologies, substitution of technologies, and involvement in a new and so far unrelated technology, which may occur through an evolutionary process.

Among the corporations, 1975 ratios of R&D to sales are positively correlated with internationalization (0.48) and size of sales (0.42) while negatively correlated with diversification (-0.28) as quantitatively approximated. Corporate strategies also emphasize growth, internationalization, and R&D, while diversification is deemphasized in half of the corporations.

R&D grows and diversifies for several reasons. There is an increased degree of external orientation and cooperation in performing R&D, and internal R&D is assuming the additional role of creating access to and possibilities for utilization of external R&D. Questions of the effects of and limits to size and growth of R&D units are discussed, but the empirical observations provide few guidelines. The concept of size of an R&D unit and the effects of size and growth appear to be a matter of organization and management more than a matter of economic scale. More pluralistic and temporary forms of organizing and managing R&D are being employed.

In discussing the findings in relation to the literature, the limited possibilities of generalizing the models for development of U.S. corporations to the corporations studied are mentioned. The connections between R&D and corporate growth are then shown to consist of both a lag between R&D and sales and a contemporary coupling through budgeting. In this way growth of R&D and sales are mutually reinforcing. In some cases R&D also lag behind investments. These relationships between growth and investments and R&D are confirmed by Schmookler. The supply- and demand-oriented theories behind the growth of technology seem to be reconcilable into a pull-push pattern when seen over an extended period of time. The long period of time involved between R&D work, and a growth in sales should also be noted. The time lag between R&D work and a significant degree of diversification is still longer in a large corporation.

Finally, corporate diversifications based on semiradical innovations are discussed in relation to external and internal ideas and impulses. The presence of contradictory cases suggests a concept of organizational permeability. This concept pertains both to the susceptibility of an organization to external ideas and impulses and to the elasticity of an organization in the event that internal ideas and impulses lead outside present product areas, as often happens. This is an evolutionary process that faces many barriers to innovation.

Appendix 3A
Main Historical Features of Corporate Growth and Diversification

Ch-Corp: Founded in the early 1870s. Based on foreign technology and domestic lag in a promising industry. Radical process invention initiated diversification but not internationalization. Continued growth and diversification parallel to technological development. Hydroelectric energy a common denominator. Severe decline in interwar years. Entrance into plastics in the mid forties. Marked diversification together with change in raw material base in the sixties. Concentration and movement into light chemicals in the seventies, together with increased internationalization.

El-Corp: Founded in the early 1890s. Based on product invention. Buildup of international organization. Continued technological developments. Diversification into several new product areas during and after World War I. Serious break caused by World War II. Rapid postwar development with diversification, growth, internationalization, and strengthening of science and technology-base.

E&F-Corp: Founded in the early 1880s. Based on product invention. Early internationalization. Strengthening of technological position. Extension of areas of application. Postwar growth and diversification. Transition from components to systems in the sixties. Concentration to main areas in the sixties through acquisitions and disinvestments.

E&S-Corp: Founded in the late 1900s. Based on product invention. Early internationalization. Integration backwards and horizontal. Temporary R&D effort. Extension of areas of application. Profitable postwar growth until the mid sixties. Late synergistic diversification.

M&M-Corp: Based on discovery of ore deposits in the mid twenties. Securing of mines and development work for utilization of ore content and subsequent diversification have been continual features. Integration forward into, among other areas, heavy chemicals in the sixties and continued diversification in the seventies. Increasing internationalization in the seventies.

P&P-Corp: Manufacturing of paper and iron dates back to the seventeenth century. Shifts in emphasis on iron and wood products until pulp began to dominate in the early twentieth century. Some diversification in the fifties. Integration forwards in the sixties. Securing of raw materials and process developments have been continual features.

Ph-Corp: Founded in 1913 in connection with a change in legislation. Initial expansion favored by World War I. Acquired 1918. Nationalized and privately reconstructed in the twenties. Break-even in the thirties. Important product innovation in the forties. Internationalization and strengthening of R&D since the fifties. No substantial diversification.

Tp-Corp: Founded in 1915 as a subsidiary to the E&S-Corporation. Start of design and production of passenger cars in the mid twenties and later on of trucks. Separation from E&S in the mid thirties. Integration backwards and other diversification. Postwar boom with heavy investments until the mid fifties. Growth of exports in the sixties and growing internationalization in the seventies. Relative growth of heavy vehicles in the seventies. Marginal diversification in the seventies.

Appendix 3B
Main Historical Features of Corporate R&D

Ch-Corp: Research laboratory started 1889. Since then ongoing R&D with varying intensity. R&D strengthened in the late 1940s. Large project failed in the 1950s. Temporary decline of in-house R&D in the early sixties with strengthening and concentration in the late sixties and the seventies.

El-Corp: Strong R&D tradition. Early product development. Research laboratory started 1914. Strengthened and diversified R&D in interwar years. Diversification during World War II and continued growth and widespread diversification since World War II.

E&F-Corp: Early product development with ongoing improvements. Application orientation in interwar years. R&D was strengthened and left the stage of materials laboratory and test shop after World War II. Further strengthening and diversification in the sixties with increased emphasis on customer processes. Short period of stagnation followed by growth and diversification of R&D in the seventies.

E&S-Corp: Materials laboratory started 1911. Successful large R&D effort around 1920. Application orientation in interwar years. Stagnation and concentration in the fifties and the early sixties. Strengthening and diversification in the late sixties and seventies.

M&M-Corp: Ongoing process developments from the beginning. Central R&D laboratory started in late 1940s. Product development oriented around utilization of by-products. Process developments stimulated integration forwards into chemistry in the early sixties. Marketing of process know-how in the seventies.

P&P-Corp: Ongoing process developments. Start of collective R&D in early 1940s. No internal product R&D until a small laboratory was started in the late sixties.

Ph-Corp: R&D started in early 1930s and strengthened during World War II. Substantial growth and diversification of product R&D initiated in the fifties and strengthened in the sixties. Continued growth but no diversification of R&D in the seventies.

Tp-Corp: Internal design from the start. Reliance on suppliers' R&D. R&D around central components successively internalized. Some diversification in the fifties and sixties. R&D strengthened in late sixties but declined shortly afterward. R&D integrated with other operations.

Appendix 3C
Cases of R&D and Diversification

E&F-Corp

After World War II the E&F-Corp experienced rapid growth within an extended range of applications. The scale of production of many customers also grew. R&D was split between machines and plant design. A small revolution came about in the technical design of one component, but otherwise the 1950s were characterized by continued growth in the technical parameters of the products, such as in motor power. The E&F-Corp developed plants for dairy and starch factories, among other things. The handling of combinations of machines became increasingly fundamental, and knowledge about the customer processes in which the machines operated grew.

During the 1960s the E&F-Corp was transformed from a component-oriented to a systems-oriented company. No specific strategic decisions were involved. Rather this transition happened successively, according to the ways in which markets and internal knowledge about customer processes developed. Studies were made of changes in customer need in marine and chemical industries, dairies, and farming all around the world, R&D grew, disinvestments were made of products and companies that did not fit in, and at the same time companies were acquired to supplement technological knowledge.

In the second half of the sixties new technologies became relevant. One reason was that an increased systems orientation brought the E&F-Corp into contact with a wider range of technologies; another was external technological change in general, and still another was problems with products and processes. Mathematics was modernized and computers came into use, automatic process control became increasingly important, and within food processing microwaves became reality. In the late sixties top management initiated an analysis of all products with respect to possible technological substitutions. Attention was paid to the risk involved in basing corporate operations on old principles and technologies, for example, separation through centrifugal forces. R&D was initiated within filters, and membrane technology and new processes came forth.

In the seventies R&D started to grow again after some years of stagnation. The competence base has been extended into areas such as biology, protein chemistry, agriculture, and electronics. The more complicated customer technologies become, the more R&D in diverse fields is needed. The nature of R&D has changed from trial-and-error experimentation to more scientifically oriented R&D. A special unit for automation technology has been created. A corporate R&D laboratory has grown up; one of its purposes is to perform R&D that does not naturally belong to a single product division or is of a long-range nature.

E&S-Corp

Soon after E&S-Corp was founded in 1907, material and quality problems led to the start of a materials laboratory. (In this respect Germany was a leading country.) Some years later E&S-Corp integrated backward into steel production in order to secure the development and supply of raw materials of high and even quality. Developments in industry in general and especially in the car industry favored E&S-Corp in its early years. A good, basic design at an early stage, a high degree of externally financed expansion (higher than the German bearing industry, for example) comprehensive management abilities plus what may be called a portion of luck in connection with World War I explain the first decade of successful corporate development.

One of the "pet ideas" of the founder was to develop bearings of universal applicability. The important and growing railway market required other design concepts with rollers instead of balls due to the high and intermittent loads involved. In the years 1918-1921 a temporary R&D effort was made for the development of roller bearings. A young engineer was given extensive power to treat the problem in its whole amplitude. A special R&D organization was set up which represented a new orientation within industrial R&D from individual inventors to R&D teams and organizations, at least in the Swedish engineering industry. (Probably the earliest example of this is Alfred Nobel's truly multinational R&D organization in the 1880s.) In 1921 the R&D staff totaled 135 persons.

The theoretical basis was weak, and much of the work had to be achieved through carefully designed experiments (cut-and-try), exploring a multitude of parallel possibilities. The specified goals of the project were finally achieved with self-adjustment as an additionally achieved feature of the design. (Self-adjustment was the basic, global improvement in the original ball bearing design of E&S-Corp giving the corporation its special competitive strength.)

In 1921-22 the unique, but not unnatural, happened—the R&D organization was dissolved. The new product was transferred to the manufacturing organization, and R&D people went back to their earlier duties. It was said that it took about ten years to achieve a truly rational production of a new product type and the design department therefore had to work fluctuatingly. (The E&S-Corp also experienced a depression in early 1920s). The diversification project then had taken three and one-half years to complete, of which one and one-half was for design, one on tests of tools and machinery, and one on experiments with mass production.

The new roller bearings gave E&S-Corp several new markets, especially railways and tramways. Sales on these markets, however, developed slowly since the customers had to carry out tests for long periods of time, before the costly and radical transition to the new type of bearings could be made.

This R&D project is a milestone in corporate development and could be characterized as the most concentrated and successful R&D effort so far in

R&D, Corporate Growth, and Diversification

corporate history. It implied a diversification within the bearing product area of enduring profitability. Achievements in coordinating and advancing production technology in the whole corporation were made in the late 1920s. In the twenties and thirties a marketing organization with application engineers was developed, and much R&D concerned new applications. After World War II, growth and profitability were good and R&D was not of primary interest.

In the sixties corporate coordination and diversification became strategic aims. A foreign-located, multinationally composed corporate R&D laboratory was inaugurated in the early seventies as well as a small, semiautonomous innovation company on corporate level. The corporate organization was strengthened and worldwide production restructured for increased coordination. Diversification is now sought through both R&D and acquisitions.

Ph-Corp

Just as it had been favored by World War I, Ph-Corp was favored by World War II; sales trebled, amounting to 20 million Sw.Cr. in 1945. An important source of profits then were sweeteners. R&D was strengthened, and a new corporate laboratory was inaugurated in 1943. The same year a highly significant event occurred when Ph-Corp acquired a newly discovered local anesthetic offered by two university researchers. After four years of industrial R&D the new pharmaceutical was introduced and caused a breakthrough with rapid internationalization of Ph-Corp. During World War II Ph-Corp also engaged in the production of antibiotics, which emerged as an important pharmaceutical area globally, due to new methods of production.

In the forties diversification outside pharmaceuticals was pushed as a strategic issue because of the perceived risk at that time of nationalization of the pharmaceutical industry. In the fifties internationalization and diversification within pharmaceuticals was also decided on, utilizing marketing synergies. This was also decided to be achieved through internal R&D at decentralized subsidiaries located close to medical universities and allowed to compete internally. In Ph-Subs an R&D organization was built up in the late fifties and external cooperation with university researchers was initiated. In the early sixties two different research lines had been established, beta-blockers and analgetics, since Ph-Subs wanted diversification in R&D in order to spread risks and also to stimulate R&D personnel. The choice of these lines very much depended on external contacts and the sequence in which these were taken rather than a strategic choice of areas and a controlled selection procedure. This kind of random element was also present in the next period of watershed years when beta-blockers were chosen for continued R&D in the mid sixties. By this time a third area of R&D was established within ulcus diseases as a result of successful sales and a subsequent decision to go in for R&D in this area.

In 1967 a new pharmaceutical, based on beta-blocking effects, was registered and introduced on the market. Although a slow starter on the market, it represented a breakthrough into a new product area. It also meant a kind of breakthrough internally in that the R&D in Ph-Subs, which earlier had been regarded with suspicion and distrust among some corporate R&D authorities, now had proven successful. After this event the need for R&D expertise grew rapidly in Ph-Subs; "research breeds research," and there was a conviction that fundamental knowledge was required for development of subsequent products. R&D volume grew in relation to turnover; this growth caused a certain resistance at the corporate level, since Ph-Corp had traditionally budgeted R&D within a narrow percentage interval of turnover.

In the late sixties it was found unexpectedly in Ph-Subs that beta-blockers had an effect on hypertension, and because of this a new and important disease area was entered. (It is, however, said to be something of a rule in pharmaceutical research that when a compound is found to have an effect within one area of indication, one has to see whether it has effects in other areas as well. An example is the local anesthetic from 1943; in 1950 it was found to have effects on disturbances of heart rhythm. This disease area was incidentally the initial target for R&D on beta-blockers in Ph-Subs.) A second generation of beta-blockers with effects on hypertension was registered and introduced on the market in 1975. A fourth research line around the central nervous system was also established in Ph-Subs in the sixties but was transferred to the parent company of Ph-Corp around 1970. In summary, then, four research lines have grown up in Ph-Subs since R&D was initiated in the late fifties. Of these four, one has proven extremely successful and raised sales in the cardio-viscular area from 2 to 145 million Sw.Cr. (13 percent to 67 percent of total turnover) in the period 1960-1975. One research line was transferred to another corporation, one was transferred internally, and in the fourth (ulcus research) there was in 1975 a wait-and-see situation, with three compounds having failed at an R&D cost of 20 million Sw.Cr. and a fourth one being tested.

As far as Ph-Corp as a whole is concerned, the strategy with decentralized R&D in subsidiaries close to medical universities was successful in some other units as well. As a result of growth and diversification of R&D in different parts of the corporation voices were raised in favor of corporate coordination in opposition to the policy of internal competition. The situation with respect to projects and licenses had run somewhat wild, and in the mid sixties a research management committee at the corporate level was created. In the late sixties R&D was profiled in the corporation, resulting in the internal transfer of a research line in Ph-Subs. In the seventies R&D has been still concentrated and consolidated within different pharmaceutical areas. Risks have to be taken on a subsidiary level and distributed on divisional and corporate levels. The corporate strategy from the fifties of internationalization and diversification within pharmaceuticals through R&D has been successful, not the least because of external cooperation with universities and other corporations in R&D and marketing.

The strategy of diversification outside pharmaceuticals has, on the other hand, partially failed. Diversification into chemical-technical and nutritional products was attempted, and the means in the latter case was R&D rather than company acquisitions. Lack of synergy with competence in pharmaceuticals and inappropriate markets for R&D-based products are among the expressed explanations for the failure of diversification into nutritional products.

References

Channon, Derek F. 1973. *The strategy and structure of British enterprise.* London: MacMillan.

Salveson, Melvin E. 1959. Long-range planning in technical industries. *Journal of Industrial Engineering,* September-October, pp. 339-346.

Schmookler, Jacob. 1966. *Invention and economic growth.* Cambridge, Mass.: Harvard University Press.

─────── . 1972. *Patents, invention, and economic change.* Data and selected essays edited by Z. Griliches and L. Hurwicz. Cambridge, Mass.: Harvard University Press.

Scott, Bruce R. 1971. Stages of corporate development, part 1. Graduate School of Business Administration, Harvard University, Boston, Mass.

4 Some Behavioral Aspects of the Coupling and Performance of R&D Groups: A Field Study

Richard T. Barth

The effective organizational integration of an organization's functions is an important issue to the R&D manager. As R&D projects and missions have become technically and administratively more complex and too large for a single group or department, functionally or otherwise specialized parties are formed in order to provide efficiencies in task performance [33] and allow each unit to deal primarily with those parts of the task environment most relevant to it [37]. With a primary focus on different parts of the environment, conflicting structural patterns may develop, often accompanied by a lack of appreciation for work flow and task interdependencies. Moreover, within each function one finds role occupants with their own predispositions, values, loyalties, interests, styles of operation, and technical expertise [31]. Various coding schemes are established as functions become more differentiated around sets of specific tasks. These coding schemes determine the amount and type of information that a given function receives, and the transformation that its members apply to it in light of the unit's systemic properties [20].

To overcome the obstacles posed by differentiation, organizational designs must provide for the enhancement of (1) accommodation between individuals in the separate functions, especially those adjacent in the flow of work, (2) required principal activities across group boundaries, that is, communication and decision making [31], along with the actual transfer of work (plans, equipment, models) such that (3) overall unity of effort results.

While a number of integrating mechanisms have been suggested and applied,[1] the specific role of their determinants needs to be better developed. There is some evidence that organizational climate may be an important determinant of communication [11, 19, 24, 32, 34] and job performance [28]. It has also been pointed out that the information exchange generated by the prevailing intergroup climate relevant to two groups may influence not only the mode or quality of joint decision making [38, 39] but also technical perfor-

---

This research was partially supported by funds provided by the Army Research Office, the National Aeronautics and Space Administration, and the National Science Foundation to the Program of Research on the Management of Research and Development at Northwestern University.

mance [2, 3]. As part of a recent study [4] the concept of organizational climate was modified to one of intergroup climate, and its impact on intergroup communication examined for sixty task-interdependent R&D groups. The objective of this chapter is to further specify the relationship between the variables of communication and decision making and, in turn, assess their impact on unity of effort (coupling) and group performance. Along with intergroup climate, this set of variables apparently has not heretofore been systematically examined in the R&D management literature.

Methodology

Data were obtained through a field study of 256 engineers-scientists and their managers (branch chiefs, section heads, laboratory directors). Interviews with 100 of the respondents and 54 managers complemented the data collected through the questionnaires. The engineer-scientists made up sixty technical work groups, each selected on the basis of the interdependence between its task activities and one specified other group (referenced group) in the same organization. Taken together, the functions of these groups encompassed pure and applied research, engineering, design, and testing activities.

Measures of Intergroup Climate

An initial pool of 125 items was drawn from instruments previously used in studies of organizational and group climate [14, 18, 24, 34]. Items were modified for applicability to intergroup climate, rewritten in terms that were meaningful to engineers and scientists, and subjected to a pilot study. These procedures yielded a final version of 68 items, each of which clearly referred to a particular aspect or facet of intergroup climate relevant to the functioning of task-interdependent technical groups. The resulting Intergroup Climate Inventory [5] allows respondents to describe, on five-point Likert scales, the intergroup climate perceived as existing between their group and the referenced group. Correlations among the items were subjected to factor analysis, using a varimax rotation. Five factors emerged, representing reasonably distinct intergroup climate dimensions, which were termed warmth/interteam spirit, risk taking, clarity, responsibility, and conformity. These results provide empirical support for the notion that the concepts of organizational and group climate can be extended to a level of analysis intermediate between group and organization: the intergroup climate relevant to two (or more) task-interdependent groups.

Perceived Communication Problems

The instrument employed to derive an indicator for this variable is based on a questionnaire originally developed in connection with phase II of Project

Hindsight [30]. The modified version used in this study was developed by Douds [10] and contains fifteen items tapping several areas of intergroup information exchange.

Joint Decision Making

Modes of joint decision making were assessed through the instrument developed by Lawrence and Lorsch [23]. Factor analysis (with varimax rotation) yielded three factors: confrontation (problem solving), smoothing, and forcing.

Group Effectiveness

The following measures of group performance were obtained through ratings provided (on seven-point Likert scales) by technical supervisors: rated "unity of effort" performance, rated innovativeness, rated effectiveness in meeting target or completion dates for tasks and projects, and rated effectiveness in reducing technical uncertainty.[2]

Interview Data

The semistructured interviews were based on a list of questions designed to obtain as accurate as possible a description of the intergroup functioning of each group with its referenced group, to learn about the recent history of projects the groups had worked on, and to obtain specific information on the nature of the coupling of the groups. The sequence of questions was structured according to the approach in which the first few items were intended to establish initial rapport through more general, less probing questions; subsequent questions were more specific [9].

Previous Results

An earlier analysis [4] evaluated the impact of intergroup climate on perceived communication problems. Multiple regression models relating the five intergroup climate variables to the variable of communication problems were fitted using stepwise regression analysis. This approach yielded a multiple correlation of $R = 0.783$ ($p < 0.001$), indicating that the variance accounted for is about 61.3 percent. The final equation revealed the dimensions pertaining to clarity, warmth, and risk taking to be significantly related with intergroup communication. A summary of the intergroup phenomena tapped by these dimensions is presented in table 4-1.

The questions that now arise are whether intergroup communication is

Table 4-1
Description of Intergroup Climate Dimensions Significantly Related to Intergroup Communication

Dimension	Description
Warmth, interteam spirit	Friendliness, warmth. Interpersonal and task interactions of a trusting (but not gullible) nature, nonmanipulative, supportive
Clarity	Interpersonal and task interactions organized rather than disorderly, confused, or chaotic; well-organized state required for the accomplishment of significant goals exists. Jobs appropriately structured; and methods, procedures, and authority sufficiently clarified.
Risk taking	Sense of riskiness and challenge reflected in the intergroup system. Group members perceive appropriately shifting balance between institutionalized ways of doing things and the calculated risk taking required for the performance of innovative, nonprogrammed tasks

related to the other principal cross-boundary activity of decision making [31] and how the pattern of the relationship of both with operational performance variables of interest to the R&D manager can be portrayed. The analysis presented in the following section is directed at this question.

Analysis

For ease of presentation, variables analyzed in this section are denoted in terms of the following mnemonics:

PCP	Perceived communication problems
CONFR	Confrontation (problem solving)
SMOOT	Smoothing
FORCI	Forcing
UNIEFF	Unity of effort (coupling)
INNOVA	Innovativeness
RETARU	Reduction of target date uncertainty
RETECU	Reduction of technical uncertainty

The relationships hypothesized were examined through the use of path analysis [21, 22]. A path model of relationships was induced through analysis of intercorrelations among variables, partial correlations of communication and decision making with performance variables, some cause-effect theorizing, and the application of program PATH.[3] The hypothesized linkages suggested in the

Coupling and Performance of R&D Groups

prior discussion are shown as elements of the path model in figure 4-1. Each path of significance is indicated by a one-way arrow leading from the determining variable to the dependent variable. The corresponding analysis of path contributions is presented in table 4-2.

Seven of the path coefficients shown in figure 4-1 are significant. For example, the path coefficient for the direct contribution of PCP to CONFR is -0.33 ($p < 0.01$). In other words, the higher the level of PCP, the less emphasis is placed on the confrontation mode in intergroup decision making. The data also reveal a negative and significant ($p < 0.01$), relationship between the forcing and smoothing modes.

Further examination of the data reveal that the impact of PCP on UNIEFF consists of a direct contribution complemented by indirect contributions based on paths modified by the three decision-making modes. A comparison of the respective total influences of PCP and CONFR on unity of effort shows that their impact is about equal in magnitude (-0.332 for PCP; 0.344 for CONFR) but opposite in direction due to the scoring scheme adopted for PCP. The impact of the two other decision-making modes is substantially less, 0.11 for SMOOT and -0.03 for FORCI. These data are supportive of the proposition that clear, timely, and open intergroup communication generally leads to a confrontation mode of intergroup decision making. Conflict (a mode of decision making other than confrontation) is more likely to develop when less specific objectives of a project are communicated to and understood by team members [36]. The combined influence of good communication and confrontation generates effective intergroup coupling. Unity of effort, in turn, has a strong direct effect on each of the performance variables. Its direct effect of 0.370 on the reduction of technical uncertainty is enhanced by a similarly strong indirect influence of 0.317 (Table 4-2) due to the groups' innovativeness. As suggested by the data, effective coupling is a consequence of the transfer of confrontation-based decisions across group boundaries. The resulting unity of effort brings about more innovative solutions to technical problems, less target date uncertainty, and less technical uncertainty. The relative impacts of communication and confrontation on these performance variables are about equal, with the largest contribution being on the reduction of technical uncertainty.

Discussion

The evidence presented here is supportive of the view that several aspects of intergroup climate are related to intergroup communication, which in turn is linked to group performance through effective decision making and coupling. Results based on the path analysis indicate that PCP and CONFR tend to relate to the performance variables as hypothesized in figure 4-1, whose sequence of variables has not heretofore been systematically examined.

Several sets of relationships are especially noteworthy. The first of these

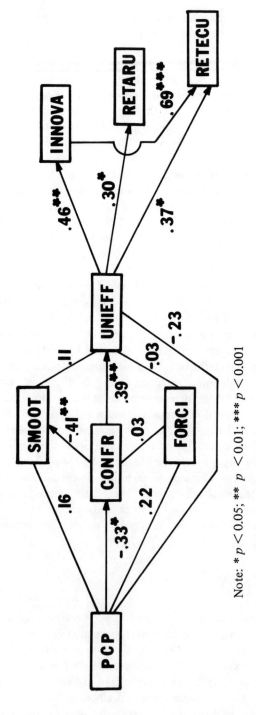

Note: * $p < 0.05$; ** $p < 0.01$; *** $p < 0.001$

Figure 4-1. Path Model of Coupling and Group Effectiveness with Path Coefficients

Table 4-2
Path Contributions to Group Performance Using Path Coefficients

	INNOVA			RETARU			RETECU		
	Direct	Indirect	Total	Direct	Indirect	Total	Direct	Indirect	Total
PCP	—	−.153	−.153	—	−.099	−.099	—	−.228	−.228
CONFR	—	.158	.158	—	.103	.103	—	.236	.236
SMOOT	—	.050	.050	—	.033	.033	—	.075	.075
FORCI	—	−.013	−.013	—	−.009	−.009	—	−.020	−.020
UNIEFF	.460	—	.460	.300	—	.300	.370	.317	.687

Note: Total contribution of PCP to UNIEFF is −.332, (direct, −.230, indirect, −.102). Total contribution of CONFR to UNIEFF is .344, (direct, .390; indirect, −.046).

refers to the relatively small influence of the conformity dimension and the preponderant influence of clarity, warmth, and risk taking on PCP. Examination of detailed interview notes revealed that a high degree of formal conformity-type features existed but was not always perceived as exacerbating intergroup communication and decision making because group members had devised informal means of overcoming them. Interviewees pointed out that it is often possible to continue work on a given set of tasks on the basis of informal interactions, while formal decisions were being processed through various levels in the organization. Collectively, they felt that this would not be possible if high degrees of warmth, clarity, responsibility, and an appropriate level of risk taking were not present to also deal with formal aspects of the organizational system (such as major decision points on large projects).

A second set concerns the highly significant negative relationship between smoothing and confrontation, and the nonsignificant relationship of the latter with forcing. As depicted in the path diagram, more emphasis on confrontation would lead to less smoothing but not necessarily less forcing. Interview data obtained from group members and their managers consistently revealed that rejection of a confrontation approach usually first led to more intensive use of smoothing for a short period of time. However, this avoidance mode was short-lived, due to the complexity and interdependence of tasks. As a result, a given group pair's decision making would take on the features of forcing, if only in anticipation of interference by management. This implies that smoothing and forcing, in that order, provided recourse to supporting or backup modes of decision making when the intergroup climate and communication were not supportive of a confrontation mode. This is in contrast to Lawrence and Lorsch's [23] finding that organizations making little use of confrontation also tend to avoid the smoothing mode and are likely to switch directly to a forcing relationship. However, the present results of a negative impact of either forcing or smoothing on performance lend support to their study.

The relationship between unity of effort and performance is also of special interest, especially due to the interplay between innovation and the reduction of technical uncertainty. More unity of effort allows for higher innovativeness. However, the generation of a more or less innovative group output involves, by definition, a greater degree of unprogrammability, technical uncertainty, and entails different information search behaviors [6] than outputs reflecting, for example, only the more or less routine application of the state of the art. The trade-off of quality of output versus meeting deadlines (both have direct budgetary and manpower implications) presents a serious dilemma to the manager. In the arena of R&D, time pressures appear to be more frequently felt than budgetary or quality demands [17]. A similar observation applies to this study, based on the interviews with group managers. A recurrent theme in these interviews concerned the viewpoint that (1) very few mission-oriented projects achieve their intended level of innovativeness (operationally defined as meeting performance or operating characteristics not attained heretofore) within the funds initially allocated; (2) usually either additional funding is provided or (3) performance goals rather than deadlines are adjusted, as the latter are closely tied to financial constraints.

One type of issue not specifically addressed in this research relates to the question whether the intergroup conflicts that occur between task-interdependent units tend to be based on task-oriented issues rather than interpersonal problems. There is some evidence that in project-oriented environments conflict is due primarily to task-focused issues [35]. Others have suggested that much of the behavior in the R&D process is highly personal, that the vitality of the process depends highly on emotions and perceptions, and that the role of individual and group values, emotions, and perceptions is crucial in determining the direction of an R&D effort [29]. Viewed in terms of the general effect of each form of conflict, Evan [13] reports a positive association between task-oriented conflict (presumably below a critical level) and team performance, and a negative relationship between performance and conflict based on interpersonal dislikes.

Practical Implications

The pragmatic reader may ask whether a number of implications may be advanced for managing the intergroup climate in order to enhance intergroup communication and performance. Our knowledge of the determinants of organizational climate is still relatively limited, and it may be fruitful to focus on variables that directly influence the organizational life of the employee. Several insights can be put forth, on the basis of interviews and a series of feedback sessions with respondents. These sessions elicited comments with respect to specific managerial actions that might be undertaken for groups reporting an unfavorable intergroup climate along with high PCP and low use of confrontation, and as part of a series of field experiments or through a program of "administrative experimentation" as defined by Campbell [8]. The following action alternatives emerged.

1. Change in group membership. Modify the composition of one or both of the groups of a given task-interdependent pair, as long as the required level of technical competence is maintained. If possible, allow for some degree of sociometric choice.
2. Change in group leader or leadership style. Transfer a member of group A to group B, although not necessarily on a permanent basis. Variations of this approach may extend to temporary reciprocal exchanges of one or more members between A and B.
3. Change in, or introduction of, a "coupling agent" or coupling system. Change the basis of performance evaluation and rewards, perhaps through

the introduction of superordinate goals, to enhance performance-reward expectancies.
4. Changes in task composition or task distribution, to provide more variety, heightened task identity, and more immediate feedback (Hackman and Lawler [16]).
5. Better recognition of the interconnectedness of time deadlines, when delineating deadlines (to forestall dysfunctional effects of ambiguous or high perceived time pressures).
6. Change in physical proximity of groups.
7. Changes in other physical aspects and modifications of the information environment.
8. Changes in reporting patterns between groups and managers.

The behavioral and intergroup climate efforts that might be anticipated as a result are several. For example, changes in spatial arrangements incorporating the "space bonds" described by Morton [27] can be expected to result in not only more frequent but more cohesive interpersonal and task interactions. The transfer of group members between groups is likely to achieve similar effects and can be expected to result in increased quality of the intergroup climate. Moreover, an increase in intergroup clarity might be accompanied by a change in the risk-taking dimension, insofar as avoidance of technical risk taking is a function of the uncertainty that group members connect with not being made fully aware of the interdependence of their task activities with those of the other group. Both these dimensions were found to be significantly related to PCP. Intergroup difficulties based on value differences may also be assuaged by this approach, especially if one group's members are attuned to science-oriented values and the other's to engineering-oriented values. Changes in group leader or leadership style, as indicated by the results reported by Meyer [26], may result in changes reflecting an increase in intergroup climate on all five dimensions identified in this study. The fourth and fifth items incorporate a motivational approach. The remainder consist of a number of procedural and organizational bonds or bridges to enhance intergroup functioning.

Notes

1. These include, for example, the use of linking, coupling, and integrating agents or units, interparty conflict management schemes, and other forms of various human, organizational, and procedural bridges [7, 15, 23, 25, 27, 31, 35].

2. The inclusion of the latter two aspects of effectiveness is based on Eyring's [12] study of sources of uncertainty, their consequences on engineering design projects, and their subsequent categorization into target and technical kinds by Abernathy and Goodman [1].

3. This computer program was written by L. Goodman (Yale University) and subsequently modified to a more general form by D.G. Morrison (Northwestern University). Briefly, the program allows the researcher to examine one or several a priori patterns of relationships and to select the one most consistent with the available correlational data. In this case, the pattern selected lends empirical support to the sequence of variables hypothesized on the basis of several studies [11, 19, 28, 39], none of which however considered the entire set of variables treated in this study. Final causal relations between the variables shown in figure 4-1 are not proved in this analysis. Given a developing rather than an established state of theory with respect to the determinants of effective coupling and group performance, this research takes a first step toward suggesting a plausible model.

References

[1] Abernathy, W.J., and Goodman, R.A. Strategies for development projects: an empirical study. Invited paper presented at the Sixteenth Annual TIMS Meeting, New York, March 28, 1969.

[2] Allen, T.J. The use of information channels in research and development proposal preparation. Working Paper No. 97-64, MIT, Sloan School of Management, 1964.

[3] Allen, T.J., Gerstenfeld, A., and Gerstberger, P.G. The problem of internal consulting in research and development organizations. Working Paper No. 139-168, MIT, Sloan School of Management, 1968.

[4] Barth, R.T. The relationship of intergroup organizational climate with communication and joint decision making between task-interdependent R&D groups. Document No. 70/34, Program of Research on the Management of Research and Development, Northwestern University, Evanston, Ill., 1976.

[5] Barth, R.T. Description and characteristics of the intergroup climate instrument for engineers and scientists. Document No. 71/67, Program of Research on the Management of Research and Development, Northwestern University, Evanston, Ill., 1971.

[6] Barth, R.T., and Vertinski, I. The effect of goal orientation and information environment on research performance: a field study. *Organizational Behavior and Human Performance* 13 (1975): 110-132.

[7] Brown, W.B. Organizational and information coupling in technological innovation. *Proceedings of the Academy of management,* Thirty-Fifth Annual Meeting, New Orleans, August 1976, pp. 478-480.

[8] Campbell, D.T. Administrative experimentation, institutional records, and nonreactive measures. In J.C. Stanley, ed., *Improving experimental design and analysis.* Chicago: Rand McNally, 1967.

[9] Cannell, C.F., and Kahn, R.L. The collection of data by interviewing. In L. Festinger and D. Katz, eds., *Research methods in the behavioral sciences.* New York: Holt, Rinehart and Winston, 1953, pp. 327-380.

[10] Douds, C.F. The effects of work-related values on communication between R&D groups. Ph.D. dissertation, Northwestern University, Evanston, Ill., 1970.

[11] Ellison, R.L., and McDonald, B.W. An investigation of organizational climate. Institute for Behavioral Research in Creativity, Greensboro, N.C., December 1968.

[12] Eyring, H.B. Some sources of uncertainty and their consequences on engineering design projects. *IEEE Transactions on Engineering Management* EM-13, no. 4 (December 1966):167-180.

[13] Evan, W.M. Conflict and performance in R&D organizations. *Industrial Management Review* 7 (1965):37-45.

[14] Friedlander, F. Performance and interactional dimensions of organizational work groups. *Journal of Applied Psychology* 50 (1966):257-265.

[15] Frohman, A.L. Stimulating research based innovation. In S.A. Johnson and J. Hartley, eds., *Proceedings of the Twenty-Seventh National Conference on the Administration of Research* (Denver, Colo.: Denver Research Institute) 1974.

[16] Hackman, J.R., and Lawler, E.E., III. Employee reactions to job characteristics. *Journal of Applied Psychology Monograph* 55 (1971):259-286.

[17] Hall, D.T., and Lawler, E.E., III. Job pressures and research performance. *American Scientist* 59 (1971):64-73.

[18] Halpin, A., and Croft, D. *The organizational climate of schools.* Chicago: University of Chicago Press, 1963.

[19] Homans, G.C. *Social behavior: its elementary forms.* New York: Harcourt, Brace and World, 1961.

[20] Katz, D., and Kahn, R.L. *The social psychology of organizations.* New York: John Wiley and Sons, 1966.

[21] Kerlinger, F.N., and Pedhazur, E.J. *Multiple regression in behavioral research.* New York: Holt, Rinehart and Winston, 1973.

[22] Land, K.C. Principles of path analysis. In E.F. Borgatta, ed., *Sociological methodology.* San Francisco: Jossey-Bass, 1969.

[23] Lawrence, P.R., and Lorsch, J.W. *Organization and environment* Homewood, Ill.: Irwin, 1969.

[24] Litwin, G.H., and Stringer, R.A., Jr. *Motivation and organizational climate.* Boston: Division of Research, Harvard Business School, 1968.

[25] Lynton, R.P. Linking an innovative subsystem into the system. *Administrative Science Quarterly* 14 (December 1969):398-416.

[26] Myer, H.H. If people fear to fail, can organizations ever succeed? *Innovation* 8 (1969):57-62.

[27] Morton, J. From research to technology. In D. Allison, ed., *The R&D game: technical men, managers, and research productivity.* Cambridge, Mass.: MIT Press, 1969.
[28] Pritchard, R.D., and Karasick, B.W. The effect of organizational climate on managerial job performance and satisfaction. *Organizational behavior and human performance* 9 (1973): 126-146.
[29] Rubenstein, A.H. Some common concepts and tentative findings from a ten-project program of research on R&D management. In M.C. Yovits, D.M. Gilford, R.H. Wilcox, E. Staveley, and H.D. Learner, eds., *Research program effectiveness.* New York: Gordon and Breach, 1966.
[30] Rubenstein, A.H. Research on proposition formation and field studies connected with project Hindsight: first technical report. Northwestern University, Program of Research on the Management of Research and Development, Report No. 66/17, 1966.
[31] Rubenstein, A.H., and Douds, C.F. A program of research on coupling relations in research and development. *IEEE Transactions on Engineering Management* EM-16 (1969): 137-143.
[32] Sells, S.B. An approach to the nature of organizational climate. In R. Tagiuri and G.H. Litwin, eds., *Organizational climate.* Boston: Division of Research, Harvard Business School, 1968.
[33] Souder, W.E. Effectiveness of nominal and interacting group decision processes for integrating R&D and marketing. *Management Science* 23, no. 6 (February 1977): 595-605.
[34] Stephenson, R.W., Gantz, B.S., and Erickson, C.E. Development of organizational climate inventories for use in R&D organizations. *IEEE Transactions on Engineering Management* EM-18 (1971): 38-50.
[35] Thamhain, H.J., and Wilemon, D.L. Conflict management in project-oriented work environments. *Proceedings of the Sixth International Meeting of the Project Management Institute.* Washington, D.C., September 18-21, 1974.
[36] Thamhain, H.J., and Wilemon, D.L. Diagnosing conflict determinants in project management. *IEEE Transactions on Engineering Management* EM-22 (1975): 35-44.
[37] Thompson, J.D. *Organizations in action.* New York: McGraw-Hill Book Company, 1967.
[38] Walton, R.E., and Dutton, J.M. The management of interdepartmental conflict: a model and review. *Administrative Science Quarterly* 14 (1969): 73-84.
[39] Walton, R.E., Dutton, J.M., and McCafferty, T.P. Organizational context and interdepartmental conflict. *Administrative Science Quarterly* 14 (1969): 522-542.

5 Organizational Factors in R&D and Technological Change: Market Failure Considerations

Almarin Phillips

This chapter analyzes some issues relating to R&D and technological change in a theoretical framework that is largely unexplored in economic literature. It addresses the influence of organizational factors internal to the firm—particularly vertical integration—and external market organizational factors on the amount and effectiveness of R&D expenditures. Adaptive response mechanisms internal and external to the firm are included in the analysis.

That this framework is not in the traditional analytic vein of industrial organization requires a brief explanation. The leading research paradigm in industrial organization for the past several decades has centered on the structure-conduct-performance approach advanced by Edward S. Mason in the 1930s.[1] No attention is given either to internal firm organization or to dynamic responses in this paradigm. Recent work by students of the so-called Chicago school has been critical of that approach.[2] While the Chicago school emphasizes the importance of the conduct and performance of individual firms on market structure and efficiency, it does so without an analytic framework dealing with the internal organizational characteristics that lead to superior performance and, indeed, without an explicit model explaining dynamic relationships among performance, conduct, and structure. Primary attention is given to the firm, to the structure and organization of firms in a market context, to their conduct and to consequent firm and market performances in both the Harvard and the Chicago traditions. If organization matters, it must be due to market organizations of which the firm is a member (cartels, tight or loose oligopolistic organizations) or to nothing more analytic than the superior organization of particular firms that underlies their superior performance.

In the last decade cogent arguments have been advanced that organizational factors internal to the firm are important in the explanation of market performances. Building on the seminal work of Ronald Coase[3] and on the foundations of organizational behavior developed by Herbert Simon and his

---

Helpful comments on an early draft of this paper were given by Oliver E. Williamson, David J. Teece, and Robert E. Olley. The discussion by participants of the Conference on R&D Management and Research Policy, International Institute of Management, Berlin (December 6-8, 1978) was also very useful. Responsibility for the final version rests, of course, with the author.

followers,[4] Oliver Williamson has provided new concepts of market failure.[5] In Williamson's view neoclassical market mechanisms are sometimes less efficient than the administratively established exchange mechanisms available through the internal organization of firms. The reasons for this, moreover, are susceptible to analysis.

Not surprisingly, there has been little effort to synthesize the "older" industrial organization approach with the "new" one. In both theory and empirical work external organizational factors (or market conditions) are not given much attention when internal organizational factors are analyzed; internal organization is virtually ignored when external factors are analyzed. Of particular importance is that work on R&D and technological change is characterized by the same dichotomy. There is a great quantity of research—most of it in the managerial tradition and most of it nonanalytic—on the internal organization of R&D laboratories and its effects on performance.[6] There is similarly a vast literature on how market structure affects technological change.[7] With little exaggeration, work of the first type fails to consider market factors; that of the second type fails to consider internal factors.[8] In general, and this is true of the industrial organization studies as well, neither type of research is couched in a conceptual framework in which a mix of performance failures and successes induces adaptive responses that alter internal and external organizations and structures through time.[9]

A General Setting

If organizational and structural factors internal to an R&D group, external to R&D but internal to the firm, and external to the firm are to be addressed, the problem is no longer one of a single focal organization. Instead, there are intersecting and interacting organizations in horizontal and hierarchical relationships. Simplification is required.

The hypothetical multiple organizational setting used here is not atypical of the real world. Let S_1 represent an organization that sells goods or services in the final market. Let M_1 represent an organization that manufactures the goods that S_1 sells or some or all of the capital goods necessary to produce the services that S_1 sells. Then add an R&D organization R_1, whose activities relate to the process and product technologies of the outputs of S_1 and to those of M_1.

Figure 5-1 shows this part of the organizational structure. It is assumed that *within* S_1, M_1, and R_1, the market mechanism is not used for transactions or for the resolution of conflicts. Each is an identifiable organization, although there are suborganizations within each of them. While recent theoretical and empirical contributions indicate that the internal organizational structures of S_1, M_1, and R_1 may affect firm performance and may themselves be altered over time because of external market forces,[10] only the internal structure of the R_1 organization is considered here. No attention is given to the organization of buyers—the final market for S_1—except that it is organizationally separate from

Market Failure Considerations

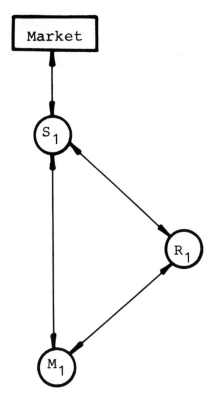

Figure 5-1. The Core Organizational Framework

S_1. The double-ended arrows show two-way interactive effects among the final market, S_1, M_1, and R_1.

A variety of organizational forms could exist within this core. Thus the alternatives could be the following:

1a. S_1, M_1, and R_1 are all divisions of one firm.
1b. S_1 and M_1 are divisions of one firm; R_1 is separate.
1c. S_1 and R_1 are divisions of one firm; M_1 is separate.
1d. M_1 and R_1 are divisions of one firm; S_1 is separate.
1e. S_1, M_1, and R_1 are all separate.

Through most of the discussion, no distinction will be made between divisions and corporate organizations under effective joint control. Being separate implies only that the market mechanism is used for transactions among the several organizations. A higher-level administrative organization controls transactions among "divisions."[11]

However, S_1, M_1, and R_1 are arranged, each may identify with other groups. Again simplifying, figure 5-2 shows an extended horizontal organizational framework. It recognizes that S_1 may be itself a part of a larger organization comprising $S_1, S_2, \ldots$, all of which sell to the final market. Along the same lines, there may be an organization of $M_1, M_2, \ldots$ and one comprising $R_1, R_2, \ldots$.[12] Conceptually, these horizontal organizations could be as follows:

2a. All S_i are divisions of one organization.
2b. Some S_i are divisions of two or more organizations.
2c. All S_i are separately owned.
2d. The same for M_i.
2e. "   "   "
2f. "   "   "
2g. The same for R_i.
2h. "   "   "
2i. "   "   "

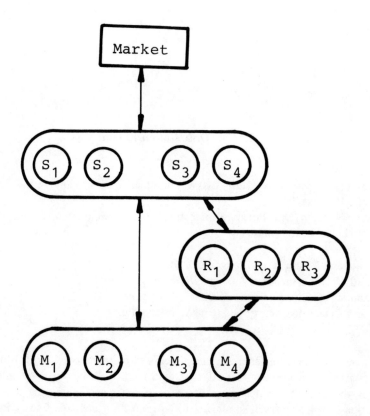

Figure 5-2. An Extended Horizontal Organizational Framework

Market Failure Considerations

The identification need not be horizontal, of course. In figure 5-3, each of the S_i, M_i, R_i sets is integrated into a single organization, but for a single case (3), no formal horizontal organization is assumed to exist among them. It clearly is possible—and observable in the real world—for many combinations of the 1a-1e, 2a-2i, and 3 to be organizationally arranged.

The inclusion of R&D groups in the settings implies that there is some value placed by some organization on accretion to knowledge. The R&D activity uses resources that must somehow be paid for, presumably by some organization that expects the value of the activity in terms of its goals to be at least equal to the resource costs. One important factor in the amount of resources required to gain new knowledge, albeit difficult to measure, is the state and rate of progress in exogenous science and technology as they relate to the achievement of goals of the focal groups of organizations.[13]

Organizational Factors Affecting R&D Efficiency

The multiorganizational alternatives illustrated in figures 5-1 to 5-3 can be compared in terms of their likely effects on R&D and technological change.

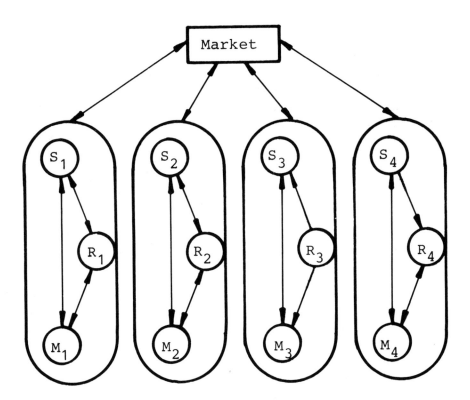

Figure 5-3. An Extended Vertical Organizational Framework

Vertically arranged organizations will be considered first. Attention then shifts to the R_1 organization and a consideration of its internal structure relative to the activities of S_1 and M_1. This leads in turn to an analysis of horizontal organizations among components of R_i organizations and their intersections with exogenous scientific and technological groups. Horizontal organizations of the S_i and M_i groups are then discussed. Incentives for organizational change and the adaptability of organizations to altering technological and market conditions are introduced throughout.

Vertical Factors

In his analysis of vertically related markets and hierarchies, Williamson stresses "bounded rationality," "information impactedness," "opportunism," and the effects of small numbers. Bounded rationality does not mean irrational behavior. Rather, it means that individuals and organizations are incapable of collecting, storing, processing, and using the complete set of information needed to make decisions in a full-scale, optimizing fashion. Information impactedness refers to asymmetries in the amount and reliability of information stored or available to different individuals or organizations. Opportunism is "self-interest with guile"—a willingness to use information strategically in one's self-interest despite possibly adverse effects on others. When all the foregoing are compounded with small numbers of parties to transactions, Williamson argues, then markets fail. Contracts that deal comprehensively with all future contingencies (bounded rationality) are difficult to formulate, the contracts tend to suboptimize to the benefit of particular parties (information impactedness, opportunism), and enforcement or amendment of contracts (due to subsequent realization of opportunistic behavior or unforeseen events) is costly. Organizational internalization of transactions is likely to be more efficient in these circumstances.

Williamson's analysis covers the 1b organization in detail. Adding R_1 as in 1a makes his arguments even more cogent. Bounded rationality characterizes the R&D activity with particular force. Indeed, the very fact that R&D is undertaken signifies that something is unknown and that knowing more is of potential but uncertain value. Still, there is no easily definable R&D output, and there are no definable measures of marginal factor contributions to guide the choice of inputs. Bounded rationality with respect to the R&D process also implies risks with respect to the ex post value of present resource use.

There are clearly asymmetries in information in the constituent parts of the 1a organization. Again, Williamson covers those relating to the S_1 and M_1 groups. The R_1 group has within itself informational impactedness because of differences in the scientific, technological, and business competence and orientation of its members. This aspect of R_1 is developed later. Beyond these internal differences there are asymmetries in information among the S_1, M_1, and

R_1 groups. With some hypole, the S_1 group knows more about the final markets, its rivals, its products, and its production processes; the M_1 group, about its products, its rivals, factor markets, and production processes; the R_1 group, about the technological environment, other research organizations, and, in a vague sense, its "production" process. Efficiency requires a linkage among them for the transfer of information.[14]

Tendencies for opportunistic behavior thus arise. There are likely to be goal differences within and among the S_1, M_1, and R_1 groups. The uncontrolled pursuit of these goals by each leads to dysfunctional performance from the point of view of the collective S_1, M_1, R_1 organization. Contracting among them has historically been viewed as an efficient way of internalizing their mutually interacting externalities.[15] The essence of the Williamson analysis, however, is that contracting is inefficient in these circumstances because of the high transaction costs involved, and the 1c-1e organizations all depend on contracting.

In this context vertical integration is an adaptive response to actual or potential market failures associated with contracting. It may not fully suppress the effects of opportunism, since bounded rationality and informational impactedness remain even with integration. Nonetheless, to the extent that integration improves the flow of communication, it also reduces informational impactedness and, hence, the opportunities for its strategic use in the attainment of subgroup goals. It may also provide incentive mechanisms that can be used administratively to overcome the attractiveness of opportunistic behavior.

The existence of an R_1 group raises two further problems which are of less significance when only the S_1 and M_1 groups appear. First, there is the "appropriability" problem. If S_1 or M_1 or a combined S_1-M_1 organization contracts with an R_1 group for development work, information must be provided to R_1 which may be of use to rivals. Enforcement of proprietary information rules by contract is typically more difficult than by administrative control. Moreover, the R_1 group has the potential for using this information *and* the knowledge it produces under contract to obtain further contracts with other S_i and M_i firms. The appropriability problem does not disappear with the integration of the R_1 group into the S_1-M_1 organization—employees may leave or communicate with others in any case—but the incentives for the R_1 group and those in it to behave opportunistically can be administratively mitigated by integration. Again, the vertical integration may be seen as a response to actual or potential failures arising from appropriability problems when contracts are used.

The second problem is probably of greater consequence than is appropriability. Karol Pelc (chapter 12) describes the difficulty with the term *range of implementation*.[16] in Pelc's words:

> Technology as a whole is considered to be a sum of specific, partly overlapping sectors.... The complexity of interrelations between members of technological family can have a positive impact on its

> development, depending on the choice of the leading technology. The same phenomenon can also exert a repressive influence, resulting in the creation of an implementation barrier due to the need to attain certain minimum levels of all sectors enabling the introduction of the required change in every single technology. Because of the nature of mutual intersectoral relations, it is necessary for the change to be made simultaneously in various sectors. The minimum level . . . required for introduction of change or for substitution of a given technology may be called an *implementation threshold.*[17]

Armour and Teece consider the same concept in terms of sharing common technologies but without addressing the implementation threshold issue.

> The existence of technologically similar production activities in various stages of an industry creates opportunities for the sharing of relevant technological innovations and refinements thereof. The existence of technological synergies increases the returns from technological endeavours, thereby inducing greater investment in R&D than might otherwise be justified. . . . If technological knowhow is to be sold via arms length transactions in the marketplace, then a number of difficulties can be anticipated. They arise from difficulties associated with trading in a commodity like technical information, the characteristics of which do not become apparent until the transaction has been completed. However, if there is sufficient precontract disclosure to assure the buyer that the information possesses great value, the "fundamental paradox" of information arises: its value to the purchaser is not known until he has the information, but then he has in effect acquired it without cost" [K. Arrow, *Essays in the Theory of Risk Bearing* (Amsterdam: North Holland, 1974)]. Since the seller is likely to be aware of the scope for opportunistic behavior by potential buyers, the seller will exercise caution and attempt to display the relevant performance characteristics while omitting other critical dimensions of knowhow. Hence, full disclosure will be avoided, and the transaction will need to take place under conditions of informational asymmetry between buyer and seller.[18]

Integration of the 1a sort helps to overcome the range of implementation, simultaneous threshold implementation, information impactedness, and opportunistic behavior difficulties described by Pelc and by Armour and Teece. One comparative advantage of 1a over any of 1b-1e is the possibility of improved communications achieved by shifting personnel among S_1, M_1, and R_1. Another is improved communications because of the use of a common coding system (language) for information transfer among vertically related sectors. Committee structures that bridge S_1, M_1, and R_1 can also be established and changed in the integrated firm.[19]

Since organizations 1b-1e all involve contracting rather than integration among at least two of the S_1, M_1, R_1, it appears that the 1a organization is

Market Failure Considerations

preferable from a market failure point of view. This generalization, however, requires some modification. Vertical integration is not itself an unambiguous term. In the present context it could range from simple common ownership of S_1, M_1, and R_1, with each operated as an independent profit center and with no coordinating, higher-level administration, to a fully centralized administration with virtually no policy decisions being made at S_1, M_1, or R_1.

Williamson notes certain limits to full vertical integration. These include internal procurement and growth biases based on factors such as organizational loyalties and goals other than profit maximization. There may also be span-of-control problems that would overtax the abilities of centralized administrators. Beyond sheer ability, the successive transfer of information from all levels within S_1, M_1, and R_1 to the central administrators and from them back to S_1, M_1, and R_1 is likely to result in communications distortions. These distortions cause administrators and managers to act not only with bounded rationality but also with faulty data due to random noise and—sometimes—opportunistically motivated nonrandom filtering errors. Finally, integration may lead to increased organizational persistence, that is, to homeostatic organizations that resist change.

These limits on the benefits from vertical integration indicate that certain of the arguments relating to the M-form hypothesis are relevant here. Negative externalities arising from full autonomy among S_1, M_1, and R_1 have to be centrally controlled. Similarly, a centrally contrived and administered incentive system can be used to minimize purposeful filtering of communications and other opportunistic behavior which is beneficial at the S_1, M_1, or R_1 levels but not at the level of the collective totality. Still, the precisely most efficient organizational form is not easily defined. Bounded rationality applies to the choice of organizational form itself. Both full centralization and full decentralization appear to produce performance failures, but the choice of the best of the intermediate forms is difficult in practice.

Organizational persistence, which Williamson suggests is a factor limiting dynamic efficiency in the inherently more centralized vertically integrated firm, exists because of the difficulty in reducing the administrative power of those possessing it due to organizational form. Paradoxically, organizational persistence also operates to resist increased centralization—that is, more administratively controlled vertical relationships—when the units are largely autonomous. Each of the separate organizations has a change-resistant internal power structure. Less centralization would improve the performance in one case; more centralization, in the other.[20] If chronic performance failures persist and the homeostatic forces of existing organization prevent their adaptation to new, more efficient forms, market forces develop for the entry of new participants with more efficient organizations.[21] Entry, that is, is another aspect of adaptive response to performance failures by existing firms.

Internal Factors

The limitations on efficiency from full vertical integration have special applicability when the internal organization of the R_1 group is considered. A clear differentiation among R&D functions is impossible because of their continuum across personnel, but the following suffices for present purposes.

R' A subgroup composed of those engaged in *exploratory research*. Exploratory research is similar to basic or pure research in that no ex ante specific application is foreseen. The research rests more on scientific paradigms than on known technologies.[22] In commercial R&D exploratory research is typically restricted to disciplines and combinations of disciplines that are perceived as being related to the product and process technologies of the enterprise.

R'' A subgroup composed of those engaged in *advanced development*. Advanced development involves perceived ex ante applications of an almost technology, "something which we know, roughly speaking, how to do [in principle] but which has not yet been put into practice."[23]

R''' A subgroup composed of those engaged in *engineering development*. Engineering development is based on a contingent ex ante commitment to an application. Alternatives are compared in terms of economic and technological criteria (costs, reliability).

R'''' A subgroup composed of those engaged in *product and marketing development*. In this development activity choices are made with respect to final configurations of products and processes, including distribution methods and modifications required in related parts of the production-distribution system.

It is important to recognize that there is no simple, causative unidirectional input-output flow among R'-R''''. In general, new products and processes cannot accurately be seen as originating in R' and then moving in coordinated fashion through the R'', R''', and R'''' activities. On the contrary, most new products and processes, at least those involving successive minor changes, are developed as a result of information flows to those in R'''' coming from others currently using existing products and processes. And the R'''' group, in turn, often passes information to the R''', R'', and R' groups.[24]

Range-of-implementation considerations require coordination between and among the R'''' and R''' groups and the S_1 and M_1 organizations, particularly when technological changes are systemic and affect the operations of both of the latter. Contracting for the R'''' and R''' activities is inefficient because of bounded rationality, unforeseen contingencies and interrelationships, opportu-

Market Failure Considerations

nism, appropriability, and small numbers. At the same time a continuous and close identification of the R'''' and R''' groups with S_1 and M_1 may lead to an analogue of internal procurement bias. Familiarity with present products and processes, plus a myopic planning horizon based on short-term goal achievements, may tend to prejudice technological choices toward minor changes that lie within the general confines of present activities.[25] Furthermore, formal horizontal organizations among the S_i, M_i, and R_i'''' and R_i''' groups, as suggested by figure 5-2, are likely to reinforce the bias toward minor changes, since rivalry internal to such organizations is suppressed.

Integration of the R', R'', R''', and R'''' groups tends to increase flows of information toward the S_1 and M_1 organizations that indicate opportunities for more radical and discontinuous change. By itself, however, integration does not fully overcome communications distortions, and the distortions may tend to favor nearly routinized, minor changes in technology by the existing organizations. That is, organizational persistence remains, and the R'''' and R''' groups may be so closely identified with the S_1 and M_1 activities that the potential contributions from R_1' and R_1'' tend to be heavily discounted.

The prevention of market failure through vertical integration thus depends on facilitating unbiased communications flows among the several groups. And, while horizontal organizations among the S_i, M_i, and R_i'''' may reinforce their internal homeostatic forces, those among the R_i' and R_i'' personnel of the several firms and with those in their scientific and technological disciplines are important sources of information about possibilities for more radical change. This presents a set of organizational dilemmas. The R_i' and R_i'' groups require a degree of isolation from the S_i, M_i, and R''' and R_i'''' in order objectively to explore technologies. Exchanges of information which at the R_i'''' and R_i''' levels raise appropriability difficulties are necessary at the R_i' and R_i'' levels in order to keep abreast of technological events. On the other hand, there must be enough identification of the exploratory research and advanced development groups with the objectives of S_1 and M_1 to suggest the types of communications flows that are useful, without at the same time restricting search because of the employment of a usefulness criteria.

Similarly, the S_1, M_1, and R''' and R'''' groups require enough awareness of developments at the R_1' and R_1'' levels to understand and use the information to advantage. The common coding and changing committee structures afforded by vertical integration aid in overcoming these dilemmas. They also create the possibility that the scientists and technologists in the R_1' and R_1'' groups will so heavily influence decisions at the R_1''', R_1'''', and S_1-M_1 levels that salient economic and market constraints are given too little attention. The relations are again nonmonotonic.

Burton Klein argues that "as the degree of uncertainty decreases, the optimal state of organization becomes more structured."[26] He uses this generalization to support the view that "open," or "random" communications within firms in high-technology industries lead to dynamic efficiency.[27] Klein's position seems more appropriate for the R'_1 and R''_1 groups and their interactions with the R'''_1, R''''_1, and S_1-M_1 groups than for the entire integrated firm. Even here, however, complete randomness (or antonomy) among the groups leads to dysfunctional performance and market failure. So do efforts to control completely the vertical relationships through administrative fiat. In terms of the short-term goals of the integrated S_1 and M_1, the R'_1 and R''_1 groups may be quite unstructured even while they are well structured to perform their function of exploring science and technology. Again, if existing firms fail to achieve an efficient organization and market failure is perceived, new firms with different organizational structures will be encouraged to enter.

Horizontal R&D Organization, "Technology Push," and "Demand Pull"

Some efforts have been made to explain technological change in terms of "technology push" and "demand pull" factors.[28] In a loose sense, the question addressed is whether necessity is the mother of invention (pull) or invention is the mother of necessity (push).

Technology, of course, does not push in the sense of mandating or forcing firms to alter their products and production processes. Technology offers opportunities to make changes. Some such opportunities are created internally through the firm's own R&D activities. Other opportunities appear from sources exogenous to the firm or, more precisely, from sources not subject to the managerial (or technological) control of the firm.[29]

In general, formal organizations among horizontally related individuals or subgroups tend to suppress some forms of strategic rivalry. That is, standardized rules and routinized behavior are established by the organization. These constrain individual behavior (opportunism) that is inconsistent with the achievement of group goals.[30] The organization, however, also forms criteria for the recognition of individual achievement and motivates those in the group to achieve through the use of accepted forms of rivalry. Organizations clarify the criteria and systematize a reward and penalty structure.[31]

Members of R'_i and R''_i groups typically belong to professional organizations in their own scientific and technologically defined disciplines. Like others who may not be employed by a market-oriented firm, they participate in such organizations because of the balance of inducements from the rewards of participation and the contributions made to the organization.[32] At the same time the R'_i and R''_i members participate in their firms, and the inducement-reward matrices of the separate organizations may not be fully congruent. The inducement to the firm (the incentive for it to contribute at least in part for membership of R'_1 and R''_1 personnel in the professional organization) comes from the potential usefulness of the knowledge it gains about exogenous science and technology. The inducement to the individual comes from the rewards received from the firm and those received from the professional organization. The former tend to be low (the firm does not contribute much to personnel engaged in professional organizations) when the firm perceives that the benefits to be derived from the discipline are low. From the point of view of those in the R'_i and R''_i groups, a high degree of congruence between the firm's reward system and the reward system of the professional association encourages high levels of contributions to both organizations.[33]

Members of the R'''_i and R''''_i groups typically belong to horizontally related R&D organizations also. While it would be incorrect to suggest that they do not simultaneously belong to disciplinary, professional associations or that the R'_i and R''_i personnel do not simultaneously participate in organizations dominated by R'''_i and R''''_i individuals, these organizations tend to be more technical than scientific or technological.[34] They often are organized with reference to a particular industry and to specific industrial applications of technology.

Opportunism, informational impactedness, and purposeful distortions of communications are not absent from the internal behavior of those in professional organizations.[35] It is, nonetheless, generally recognized that the free exchange of information is beneficial to developments in the field and even more strongly recognized that publication, especially first publication, is a primary reward.[36] It is from this largely free informational interchange that high-technology firms receive potential benefits. Further, because the transition path from scientific and technological discovery to a useful new product or process is usually uncertain, time-consuming, and costly, the release of information based on R'_1 or R''_1 work may not seriously jeopardize S_1 or M_1 goals.[37] When it may cause jeopardy, release is selective, and the R'_i and R''_i personnel find themselves in a conflicting system of rewards and constraints.

Information release at the R'''_i and R''''_i level of interfirm R&D organizations depends on the nature of the organizations and the types of rivalry among the firms. If the organization does not suppress the strategic and opportunistic use of product and process innovations, release will be highly selective because individual firm goals are the more important ones. On the other hand, some interfirm R&D organizations (patent pools) coordinate and, because of controlled use consonant with interfirm goals, encourage the mutual exchange of R&D results.

In this context "technology push" could be used to describe an innovation that has clear and direct antecedents in discoveries or developments by those not motivated by the goals of a market-oriented firm at the time of the discovery or development.[38] Obviously the training and primary motivation and goals of the discoverer need not vary depending on whether he is or is not employed by a firm. The corporate employee may be as motivated by rewards from the horizontally related professional organizations in which he participates as are others in the field. Indeed, the discoverer working within the firm may have just as much difficulty in selling his ideas to those in the R_1''', R_1'''', and S_1 or M_1 groups as an outsider. Just as obviously, an outsider may have as strong an economic motivation as the corporate employee. In either case, push can be vaguely seen.

"Demand pull" innovations, on the other hand, could describe those whose antecedents are clearly derived from fulfilling perceived needs. There is a popularized myth that a pull from final consumers—from the market in figures 5-1 to 5-3—induces most R&D and technological change. Casual observation suggests that this is rarely true. In many cases consumers have resisted product changes, particularly when the changes are radical.

There is still an important pull factor. It comes from demands made by the S_i's directly on the M_i's, directly on the R_i's, and indirectly on the R_i's through demands on the M_i's, or from the M_i or R_i groups, for lower costs, improved qualities, or new products to meet the goals of the firm. Any of the S_i's, M_i's, or R_i's of course, may perceive a latent demand for a new or improved product in the final market, but the explicit expression of this need—especially when the change is radical and not one of minor improvements obvious to all—usually comes from sources internal to the firm. Whether the perceptions of latent demand in fact originate from events in science and technology or from market phenomena is an unanswerable question of metaphysics.

Another less metaphysical demand pull factor comes from range-of-implementation considerations. Changes at one stage, whether originating from push or pull, require changes at another. Thus within and among the S_i and M_i, technological interrelationships do induce technological change in a derived demand sense whatever the source of the primary change.

Organizational analysis indicates no clear dichotomy between demand-pull and technology-push innovations. Still, the concepts suggest factors involved in market failure that an efficiently organized firm should avoid. Unused technological opportunities signal failure in a social sense. They may signal failure for the firm if another successfully utilizes the opportunities. If technology is pushing, the efficient economic organization is one that is able and impelled to respond to the exogenous developments.

Similarly, if there is a demand that can be satisfied only through technological change, the efficient economic organization is one that can and will respond. Whether the demand is expressed in the final market, perceived by S_i, M_i, or R_i groups, or arises from "gaps" in interrelated stages of technological change, organizational response is required for efficiency.

Intersections of Vertical and Horizontal Organizations

The push-pull dichotomy is further obscured when internal organizational factors include the intersectional loci among organizations. The efficiency-

Market Failure Considerations

inducing aspects of vertical integration in the process of technological change, however, become more evident.

The horizontal organizations that include the R'_i, R''_i, R'''_i, and R''''_i personnel serve as communications systems for their members. Each of the organizations provides message channels, has its own common language coding, and receives, processes, and transmits information from and to its members. It also listens to and processes information from its external environment. The organizational communication systems reduce uncertainty for members and increase the bounds of cognitive rationality. The R_i organizations thus become antennas and data processors that keep members informed about scientific and technological developments and provide input information for the R&D work of the members.

While they have not been emphasized to this point, there are also horizontal organizations among the S_i's and M_i's (figure 5-2). In most industries, there are many such organizations, reflecting coalescences of the various subgroups within the S_i's and R_i's. They provide the same sort of communication systems for their members as do the R_i organizations, and their members are also members of the several firms in the hierarchically structured production process from the M_i's through the S_i's to the market.

If the R_1's were not integrated with the S_1 and M_1, the output of the R_i horizontal organizations would have to be transmitted to S_1 and M_1 by contract or common media. So, too, would information from M_1 to S_1, and vice versa. If R_1 were integrated with S_1 but not M_1, one link of the information transmission system would be organizationally internalized but the others would not. Without the 1a type of organization, the failures inherent in contracted information transfers would obtain.

It is not just communications at each of the levels from and to its horizontal organization that are involved. As information is received at one level, say R_1, it is processed at that level, decoded from the language of that level, and encoded to the language pertinent to the goals of the firm. The communication system of the vertically integrated firm hence receives, processes, and transmits information from and to all levels in a manner more efficient than would combinations of horizontal and vertical contractual systems. The latter, indeed, might increase uncertainty and increase opportunism because of the complexity of sets of bilateral contracts.

Klein maintains that the existence of rivalry, or competition, or threats to achievements of some kind is a necessary (but not sufficient) condition for dynamic efficiency.[39] This is apparently so, but the necessary condition need not be represented by those assumed in the static theory of perfect competition in economics. An M_1 group may feel a threat from information received horizontally or vertically. The threat may be felt from information involving price, costs, input supply conditions, the general economic climate, government regulatory actions, or minor or major technological changes. The same is true at S_1 and R_1, although the dimensions of goal achievement being threatened may vary among them. The important point about vertical integration of the 1b kind is that threats at one level, especially when coded to the common language of the several stages, induce responses at all stages. With contracts, threats at one stage may induce dysfunctional opportunism rather than efficient coalescence. Thus if the perceived threat comes from a pull factor recognized by S_1, the R_1 and M_1 groups can attempt to marshall the appropriate technology to meet the threat. If the perceived threat comes from a push factor recognized by R_1, the S_1 and M_1 groups may be convinced that they should alter their products and processes. Push and pull are working in both cases. And with integration, range-of-implementation problems are at least partially overcome by a complementary and internalized range of communications.

Conclusion: The Adaptive Response Mechanisms and Efficiency

Market failure may occur, it appears, because of numerous inefficiencies in organizational forms. To say that the 1a organization is more efficient than any of the 1b-1e has elemental truth, but the 1a organization may fail because of too little or too much centralization. In the context of technological change, it may also fail because too few or too many resources are committed to the R_1 activity, because the mix of R_1'-R_1''' is inappropriate, because the contributions to the R_i horizontal organizations are too great or too small, or because the processing of data received from the R_i horizontal organizations is incomplete or inaccurate. Similarly, too much or too little may be contributed to or received from S_i and M_i organizations, and threats that should be recognized from these sources may go unheeded. Finally, the horizontal organizations may so reduce internal rivalry that none of their members perceives threat.

Dynamic characteristics of adaptive responses to inefficiencies are at least as important as the comparative static properties of alternative organizational forms. Williamson has shown how performance failures commonly shared by horizontally related firms lead to increased communication and raised levels of interfirm adherence as a corrective mechanism.[40] This is an organizational adaptive response at interfirm level. Phillips, followed by Nelson and Winter, shows that failures of the sort described—especially failures of firms to utilize efficiently the opportunities afforded by science and technology—cause forces to arise that alter the S_i structure.[41] Successful firms replace the less successful, but over time once successful firms may themselves become unsuccessful. While emphasizing adaptation, these studies do not address responses in the form of internal organizational change.

There is growing evidence that the post-World War II shifts from U-form to M-form internal organization was a response to perceived threats to performance.[42] The same is probably true of vertical integration of the 1b kind. When an active and related science or technology appears, successful firms in the high-technology industry are likely to be those that respond by adding an R_1 group, as in 1a, with intersecting organizational ties with R'_i-R''''_i horizontal organizations. Surges in science and technology related to an industry's activities tend to be accompanied by such responses, leading to industries having structures such as those in figure 5-3, even though they may have been structured differently prior to the surge.[43] The structure in figure 5-3 is, however, oversimplified since each of the successful firms in it will require intersections with R'_i-R''''_i organizations to adapt efficiency to the technological surge.

Structural transitions are likely to involve the deaths of some organizations and births of others. At the extreme, all the firms existing prior to a technological surge may fail, being creatively destroyed in the perennial gale of technological change.[44] This occurs when the existing organizations are unresponsive, fail to adapt, or make errors in the process of adaptation. The vertically integrated firm is not immune to such failures even while vertical integration may be another necessary condition to prevent failure.

Notes

1. For an overview of the Mason approach, and in addition his own works, see J.W. Markham and G.F. Papanek, eds., *Industrial Organization and Economic Development: Essays in Honor of Edward S. Mason* (Boston: Houghton Mifflin, 1970).

2. The Chicago school is illustrated, among other places, in R.A. Posner, *Economic Analysis of Law,* Boston: Little, Brown (1972) and *Antitrust Law: An Economic Perspective* (Chicago: University of Chicago Press, 1976). For contrasts between the Harvard and Chicago views, see H. Goldschmid, M. Mann, and J. Weston eds., *Industrial Concentration: The New Learning* (Boston: Little, Brown, 1974), and "Symposium on Antitrust Law and Economics," *University of Pennsylvania Law Review* 127 (April 1979):918-1140.

3. R.H. Coase, "The Nature of the Firm," *Economica* 4 (November 1937):386-405.

4. In particular, J.G. March and H.A. Simon, *Organizations* (New York: John Wiley & Sons, 1965); R.M. Cyert and J.G. March, *A Behavioral Theory of the Firm* (Englewood Cliffs, N.J.: Prentice-Hall, 1963).

5. O.E. Williamson, *Markets and Hierarchies: Analysis and Antitrust Implications* (New York: The Free Press, 1975).

6. See, for example, P.H. Francis, *Principles of R&D Management* (New York: ANACOM, 1977); D. Allison, ed., *The R&D Game: Technical Men, Managers, and Research Productivity* (Cambridge, Mass.: The MIT Press, 1969); M.C. Yovits, D.M. Gilford, R.H. Wilcox, E. Staveley, and H.D. Learner, eds.,

Research Program Effectiveness (New York: Gordon & Breach, 1966); W.E. Souder, "Effectiveness of Nominal and Interacting Group Decision Processes for Integrating R&D and Marketing," *Management Science* 23 (February 1977):596-605; J.M. Allen "A Survey into the R&D Evaluation and Control Procedures Currently Used in Industry," *Journal of Industrial Economics* 18 (April 1970):161-181; A Gerstenfeld, *Effective Management of Research and Development* (Reading, Mass.: Addison Wesley, 1970).

7. For a survey, see E. Mansfield, *The Economics of Technological Change* (New York: W.W. Norton, 1968); and M. Kamien and N. Schwartz, "Market Structure and Innovation: A Survey," *Journal of Economic Literature* 13 (March 1975):1-37.

8. B. Klein, *Dynamic Economics* (Cambridge, Mass.: Harvard University Press, 1977) is an interesting attempt to consider both.

9. Building partially on A. Phillips, *Technology and Market Structure: A Study of the Aircraft Industry* (Lexington, Mass.: D.C. Heath, 1972), R.R. Nelson and S.G. Winter, "Forces Generating and Limiting Concentration under Schumpeterian Competition," *Bell Journal of Economics* 9 (Autumn 1978):524-548, illustrates an attempt in this direction. It uses a simulation model to trace changes in market structure over time.

10. Cyert and March, *Behavioral Theory of the Firm*, describe the internal behavioral aspects of firm conduct with assumptions of bounded rationality, uncertainty, and goal conflicts. O.E. Williamson, *The Economics of Discretionary Behavior: Managerial Objectives in the Theory of the Firm* (Englewood Cliffs, N.J.: Prentice-Hall, 1964); R. Marris, *The Economic Theory of "Managerial" Capitalism* (London: Macmillan, 1964); and W.J. Baumol, *Business Behavior, Value, and Growth* (New York: Harcourt Brace & World, 1959) all investigate firm conduct when subgroup goals other than profit-maximization appear. More relevant for analysis of alternative internal organizational forms, however, are O.E. Williamson's *Corporate Control and Business Behavior* (Englewood Cliffs, N.J.: Prentice-Hall, 1972) and the empirical tests of the M-form hypothesis by H.O. Armour and D.J. Teece, "Organizational Structure and Economic Performance: A Test of the Multidivisional Hypothesis," *Bell Journal of Economics* 9 (Spring 1978): 106-122, and P. Steer and J. Cable, "Internal Organization and Profit: An Empirical Analysis of Large U.K. Companies," *Journal of Industrial Economics*, 27 (September 1978):13-30. Steer and Cable suggest that internal organizational change tends to occur as an efficiency-inducing response to external pressures on performance.

11. Organizational persistence may tend to be greater in an organization of legally defined divisions than in one composed of separate corporations.

12. In this context, an organization need not be a legally recognized entity. It may even be an illegal entity depending on its activities. Briefly, an organization is a group capable of communicating, verbally and nonverbally, with one another, having one or more collectively shared goals and (actually or

potentially) conflicting subgoals that can result in conflict situations. See A. Phillips, *Market Structure, Organizations, and Performance* (Cambridge, Mass.: Harvard University Press, 1962), especially chapter 2, "A Theory of Interfirm Organization."

13. A discussion of this factor is included later. For background see Phillips's, *Technology and Market Structure*; R.R. Nelson, M.J. Peck, and E.D. Kalachek, *Technology, Economic Growth, and Public Policy* (Washington D.C.: The Brookings Institution, 1967):34-43; F.M. Scherer, "Firm Size, Market Structure, Opportunity, and the Output of Potential Inventions," *American Economic Review* 55 (December 1965):1098-1121; M. Kamien and N. Schwartz, "Market Structure"; and Nelson and Winter, "Forces."

14. On this point, see H.O. Armour and D.J. Teece, "Vertical Integration and Technological Innovation," Research Paper No. 506, Graduate School of Business, Stanford University (August 1979). As background, see P.R. Lawrence and J.W. Lorsch, "Differentiation and Integration in Complex Organizations," *Administrative Science Quarterly* 12 (June 1967):1-47; E. Mansfield and S. Wagner, "Organizational and Strategic Factors Associated with Probabilities of Success in Industrial R&D," *Journal of Business* 48 (April 1975):179-198.

15. R.H. Coase, "The Problem of Social Cost," *Journal of Law and Economics* 3 (October 1960):1-44.

16. Pelc indicates that the concept has also been called "a family of technologies." See also J.D. Thompson, *Organizations in Action* (New York: McGraw Hill, 1967), pp. 14-16 where the term *long-linked technology* is used. Thompson does not, however, treat long-linked *changes* in technology.

17. K.I. Pelc in chapter 12 of this volume, (emphasis in original).

18. H.O. Armour and D.J. Teece, "Vertical Integration and Technological Innovation," pp. 2-3.

19. Ibid., pp. 3-4.

20. The so-called Hawthorne effect, based on observations that any organizational change is better than none, may be explained in part by this nonmonotonic relationship.

21. On the formation of new organizations, see A.L. Stinchcombe, "Social Structure and Organizations," in J.G. March, ed., *Handbook of Organizations* (Chicago: Rand McNally, 1964). A treatment more consonant with the arguments given here is A. Phillips, "An Attempt to Synthesize Some Theories of the Firm," in A. Phillips and O.E. Williamson, eds., *Prices: Issues in Theory, Practice, and Public Policy* (Philadelphia: University of Pennsylvania Press, 1967). New organizations include those formed by components of existing groups in the face of organizational disequilibrium of the Bernard-Simon type. See March and Simon, *Organizations*, pp. 83-111.

22. See Michael Fores in chapter 13 of this volume.

23. For a discussion of "almost technology" in relation to science and technology, see Derek de Solla Price, chapter 14 of this volume.

24. Derek de Solla Price, chapter 14, emphasizes this point. He observes, as have others, that "thermodynamics owed much more to the steam engine than ever the steam engine owed to thermodynamics."

25. On this line Joseph Schumpeter wrote, "It is not only objectively more difficult to do something new than what is familiar and tested by experience, but the individual feels reluctance to do it and would do so even if objective difficulties did not exist. This is so in all fields. Thought turns again and again into the more accustomed track even if it has become unsuitable and the more suitable innovation in itself presents no particular difficulties." *The Theory of Economic Development* (Cambridge, Mass.: Harvard University Press, 1934), p. 86. The same point is emphasized by Thomas S. Kuhn, *The Structure of Scientific Revolutions* (Chicago: University of Chicago Press, 1962), and by Klein, *Dynamic Economics.*

26. Klein, *Dynamic Economics*, p. 164.

27. Ibid., pp. 154-175.

28. For a discussion, see E. Mansfield, J. Rapaport, A. Romeo, E. Villani, S. Wagner, and F. Husic, *The Production and Application of New Industrial Technology* (New York: W.W. Norton, 1977), pp. 26-28.

29. Phillips, *Technology and Market Structure*; Nelson and Winter, "Forces."

30. See Phillips, *Market Structure, Organization, and Performance,* and O.E. Williamson, "A Dynamic Theory of Interfirm Behavior," *Quarterly Journal of Economics* 79 (November 1965):579-607, for analysis of horizontally related organizations among firms. More generally, see March and Simon, *Organizations,* pp. 34-47.

31. March and Simon, *Organizations,* pp. 52-82. Thompson, *Organizations in Action,* p. 120-121, stresses organizational factors that make a reward system ineffective, even without an obvious system of penalties.

32. March and Simon, *Organizations,* pp. 84-93.

33. This helps to clarify several points. First, it indicates why a distinction is made between basic or pure research and exploratory research. Firms are selective in the disciplines they support. Second, it indicates in a general way the difference between a high-technology industry and other industries. High-technology refers to those industries that can and do make use of scientific and technological disciplines and support R_i' and R_i'' activities. Third, together with other parts of the discussion, it explains why most R&D expenditures are at the R_i''' and R_i'''' levels. The contributions of the latter to the goals of the S_i and M_i organizations are more obvious and, in most cases, greater than are those of the R_i' and R_i'' groups.

34. See Fores, chapter 13 of this volume, for a scientist's lament that the distinction between "techniks" (technicians) and scientists and technologists has been obscured. Price, chapter 14, seems to carry this argument to the point that nothing done in industry should be called science.

35. See, for example, the practices described in J. Watson's, *The Double Helix* (New York: Athenium, 1968). C.P. Snow, *The Affair* (London: Macmillan, 1960) indicates similar behavior.

36. See chapter 14 of this book. Note also that many scientific journals publish submission dates for accepted articles.

37. As an example, the availability of patent information is often an inadequate base on which to proceed with production. It may provide a description of the results of basic or exploratory research, but not the results of development work. Standard Oil had the I.G. Farben patent for synthetic rubber but found it of no value without access to Farben's technical know-how. See J. Barken, *The Crime and Punishment of I.G. Farben* (New York: The Free Press, 1978), chapter 4. Decisions are often made not to patent discoveries for the same reasons and sometimes with the same effects on research personnel.

38. The tracing of the sources of particular innovations is very difficult. (See chapter 14.) J. Jewkes, D. Sawers, and R. Stillerman, *The Sources of Inventions* 2nd ed. (New York: W.W. Norton, 1969) attempt such a tracing to distinguish between the role of the individual inventor and invention by industrial R&D laboratories, but the ascription of source is not always unambiguous.

39. Klein, *Dynamic Economics.*

40. Williamson, "A Dynamic Theory of Interfirm Behavior."

41. Phillips, *Technology and Market Structure;* Nelson and Winter, "Forces."

42. See Armour and Teece, "Organizational Structure" and Steer and Cable, "Internal Organization."

43. Armour and Teece, "Vertical Integration and Technological Innovation," provide empirical evidence that vertical integration has a positive effect on basic research, applied research, and development expenditures. The argument here is that *some* of the explained variance may be due to the opposite directional causation. That is, scientific opportunities give rise to R&D and to vertical integration in order to use the opportunities efficiently.

44. This description is, of course, Schumpeter's. See J.A. Schumpeter, *Capitalism, Socialism, and Democracy* (New York: Harper & Row, 1942).

Part II
Technological Change

The chapters in this section seek an understanding of the process of technological change in its own right. Their focus is on three problem areas concerning the origin, adoption, and the development of new techniques.

The chapter by Josef Steindl offers a new philosophical and analytical perspective on technology. Technical progress is essentially an interplay of certain internal factors and occasional influences from outside. On the one hand, the process of discovery exhibits a distinctively inner logic of its own. On the other hand, it is characterized by certain abrupt changes from outside. There is an element of chance to it. One of the most important internal factors in the process of technical change is acquisition of relevant know-how over the course of time, or what might be called a process of learning in short.

There are two types of learning. First, there is the short-term learning giving rise to numerous minor innovations on a continuous basis. Second, there is the long-term learning that makes the economies of scale possible from major new changes in the productive apparatus over the course of time. The latter are in turn a consequence of the expansion of the market. However, learning by itself probably cannot account for all the observed growth of productivity. One must therefore look for the role of innovations, in particular the radical innovation, from outside. In this endeavor, network analysis and systematics can serve as useful methodological techniques.

The chapter by Cantley and Glagolev investigates the role of scale factors, which is distinct from but closely related to the role of learning in the process of technological innovation. The problem of determining the appropriate scale of technology arises in many different contexts—supertankers, collieries, industrial plants, R&D organizations. However, a general framework for scale-related decisions is lacking. In many cases the only guide is the lessons learned from mistakes committed in other areas in the past. In hitherto well-understood areas new problems have arisen as a result of unprecedented growth in the scale of technology. A prime example of this is electricity generation, where significant diseconomies of scale appear to have arisen in recent years. A consideration of some of these problems has led Cantley and Glagolev to conclude that it is useful to distinguish between several different levels of scale, for example, the size of a single piece of physical equipment, plant size. Regardless of the level chosen, however, it seems that scale-related decisions are best pursued in the larger context of an organizational setting and with an awareness of the dynamic process of learning.

The chapter by Davies is concerned with the adoption of new techniques over the course of time. It is based on an investigation of firms responsible for innovating eighteen new techniques in manufacturing in the United Kingdom.

Firm size emerges as one main determinant of the diffusion of technology. Specifically, the probability of adoption of a new technique is found to be a monitonically rising S-shaped function of firm size. This does not necessarily mean that large firms are more progressive. Rather, technical reasons, such as economies of scale, suggest that larger firms might derive greater returns from investment in new techniques. Further, while larger firms tend to be quicker in adopting new techniques already introduced by other firms in the industry, they themselves tend not to be innovators; they avoid instigating the innovation in the first place. The evidence for the relationship between diffusion speed and industrial concentration is, however, mixed. There is some indication that diffusion tends to be slower in industries characterized by higher concentration (large inequalities in firm size), but occasionally an inverse relationship between diffusion speed and degree of competition is found.

The chapter by Sahal presents a comprehensive theory of development of new techniques which is found to hold in a variety of cases of technological change in agriculture, transportation, and the computer industry. The study has two basic propositions: one, the learning-by-doing hypothesis is that technological change grows out of experience. The other, the specialization-via-scale hypothesis is that technology is a function of the scale of the system into which it is embedded. These two propositions complement each other in that the former is concerned with the temporal while the latter is concerned with the spatial aspects of the phenomenon.

Four principal conclusions emerge from the study. First, the role of learning in the process of technological innovation is much more important than hitherto recognized. The scope of learning is not confined to the use of a given technology. The origin of a technology itself is also conditional on learning. Second, different types of technology vary greatly in the extent that advances result from learning. Third, in certain cases, such as transportation technology, learning is found to play as important a role as does scale factor in the process of technological innovation. Fourth, in many instances, learning takes place largely in the capital-producing rather than the capital-using sector of the economy. This is not to say that the converse cannot occur. It is rather that the available evidence strongly points to the role of (cumulated) gross investment as an essential factor in facilitating the process of technological innovation. This confirms what has long been suspected but never adequately proved, that to a certain extent, all investment has ipso facto the character of investment of R&D.

These results have important policy implications. First, technological innovation is not just a matter of adequate investment in the R&D laboratories. It is as much a function of the growth of capital-producing sector (machine-tool industry) and small-scale workshops engaged in seemingly routine work. Second, in many instances, there may be little, if any, justification for tariffs and other forms of subsidies aimed at protecting infant industries. This is because any protection offered to the capital-using sector amounts to penalizing the capital-producing sector of the economy, insofar as learning may take place largely in case of the latter rather than the former. Instead, an appropriate policy is to eliminate any obstacles in the way of industrywide diffusion of know-how in the capital-producing sector. Finally, a central feature of the learning process is that the knowledge acquired from operating on a large scale is not foregone if the scale is subsequently reduced. That is, the benefits from learning are of an irreversible nature. Moreover, the role of learning is found to be at least comparable to changes in scale. Thus large-scale technology is not indispensable. The acquisition of production skills is a relevant alternative.

6

Technical Progress and Evolution

Josef Steindl

This chapter is concerned with two ideas: first, the analogy between technical progress and biological evolution and, second, the interaction of science and society (of which the economy is only a part).

The analogy is based on the view that technical progress is a learning process and so is evolution in biology. The questions, Who is learning, and What is being learned, are ultimately rather puzzling in both cases, but we shall for the moment concentrate on the commonsense view that there is a process of trial and error, in which one solution is retained and the others are discarded. The object of learning in both cases is technique, for example, techniques of locomotion in the case of animals, techniques of transport in the other case. The criteria for selection are provided by the environment which in the case of human techniques includes not only nature but also society and the economy.

The techniques are embodied in species; in fact, a species can be regarded as a particular combination of techniques. In a similar way, the human techniques in certain combinations are embodied in the commodities made with their help; they are also embodied in equipment; and they are embodied in industries, as well as in trades (skills). To some extent, therefore, the evolution of techniques is reflected in the development of new industries, commodities, and trades.

In the emergence of new techniques there is an element of chance; at least its timing is a random factor. This applies to the appearance of mutations. It is also true for the discoveries and inventions.

New techniques, however, have a tendency to spread to new fields: That is, they combine with a variety of other techniques. In this way, whole classes or families of animals emerge, all based on one technique but in a variety of combinations with others, like variations on a theme. In the same way an industrial technique may gradually spread to various applications, to various industries. This may be called diffusion of higher order.

A technique, on the basis of experience gathered in its application, may give rise to a new and better one. In nature this happens gradually, but in industry a major improvement may appear like a jump after a certain time. Here we come up against a great difference between the two cases. Mixture, the meeting of various combinations, is fruitful in any case. In nature it happens in the course of meiosis and fertilization; but this recombination of elements is confined to the genetical pool of a certain species. In the human case, however, ideas from the most varied fields, from anywhere, may in principle meet and give rise to something new. The field on the cultural border is the most fertile for original innovation (Ayres 1944).

There is another way in which techniques may tend to multiply. They may create a new environment and this may give rise to new techniques (or combinations of them). The environment is continuously being created and recreated by evolution itself. In this there is again no difference between nature and industry. The environment for an animal is largely constituted by other animals and plants. Again, the environment which determines the success of a method is constituted by the existing industrial techniques, the economy, and the society, and these have been created largely by the technical evolution of the past.

Thus evolution seems to be a self-perpetuating process. Yet our way of thinking and looking at things makes us see the environment and the techniques as contrasting, two open systems acting on each other. This leads us to the second of the two themes of this chapter.

The development of science is pushed along by two different forces. One comes from inside, from the inner logic of science; the other comes from outside, from society. This can be illustrated very well in the case of mathematics. In mathematics there is an inherent tendency for generalization, for extending the field of application of concepts, leading, for example, to negative, irrational, complex numbers, and from there to the theory of functions. Another inherent aim is to establish relations between seemingly unrelated problems so as to unite them under the common cover of one concept or theory. Another illustration of this inherent force is David Hilbert's list of unsolved problems, a long-term program of research arising from the tensions within the subject. At the same time it is difficult to deny that mathematics received strong impulses from technical tasks such as surveying and navigation and from scientific subjects such as astronomy and mechanics. Newton developed the calculus because he needed it for his mechanics; Laplace and Poincaré were inspired by astronomy. Probabilistics comes from the gamblers; practically all elementary problems in stochastic processes were tasks set by scientists or technicians.

The same combination of an inner force and outside influence is found in the sciences proper. The establishment of a system of chemical elements and the search for those not yet found is nearly parallel to the example given for mathematics. There is an inherent tendency to establish an order and to unify observations under some all-embracing concept.

Hardly a word need be lost on the importance of the production technique and the economy for scientific evolution. It was of two kinds. The society set tasks, for example the calendar reform; on the other hand, production technique preceded scientific development, as in the steam engine; or it was a precondition for an advance insofar as it provided instruments and apparatus. This is strikingly true of the breakthroughs in astronomy in the last decades.

There is a set of influences that I have not yet mentioned: the irrational.

Progress Functions and Scale Effects

Learning by doing has come within the scope of economists' concepts only in recent decades, largely through the progress function. This has been distilled from the data collected in the U.S. aircraft industry since the last war. The direct labor cost of the Nth airframe, c_N, is a decreasing function of N, the cumulative output of the model.

$$c_N = aN^{-b}, \quad 0 < b < 1. \tag{6.1}$$

This is called the progress function or the learning function. The parameter b, which indicates the learning effect of an additional output, is of the order of one-third in the aircraft industry.

We may write expression 6.1 as an exponential with $\ln N$ as a variable in the exponent. This is to suggest visibly that cumulative output (or rather its ln) can be regarded as a measure of time (operational time). From 6.1 we obtain the total cumulative direct labor cost C_N of the output N.

$$C_N = a(1-b)^{-1} N^{1-b}, \quad 0 < b < 1. \tag{6.2}$$

Let us, for the sake of argument, assume that the rate of input of direct labor is constant in time so that the cumulative labor cost is Ct. Inserting that for C_N in 6.2 we can calculate the cumulative output as a function of time.

$$N_t = \left[\frac{(1-b)\,Ct}{a}\right]^{\frac{1}{(1-b)}}, \quad 0 < b < 1. \tag{6.3}$$

It follows that the output per unit of time N_t/t of a given labor force is a function of $t^{\frac{b}{(1-b)}}$. If $b < \frac{1}{2}$ (as it apparently is empirically) then the output per unit of time increases less and less as time goes on. It is not surprising that the learning effect should peter out after a time, since we are dealing with the output of a particular model with given equipment in a given firm. The possible improvements of organization of work become largely exhausted after a time. Therefore a progress function of this type could not explain a more or less steady growth of productivity in the economy as a whole; nor could it explain its prevailing magnitude in relation to the growth of output. Productivity growth in advanced countries in the postwar decades usually accounted for much more than half of the growth of output, which would be difficult to reconcile with the practical values of b. Productivity is

$$N/C_N = \left[(1-b)/a\right]^{\frac{1}{(1-b)}} N^{\frac{b}{(1-b)}}. \tag{6.4}$$

If output rises exponentially at a rate ω, then inserting this in 6.4 gives

$$N/C_N = k \exp\left\{\frac{\omega b t}{(1-b)}\right\},$$

that is, productivity grows at the rate of $\omega b/(1-b)$ which means that its growth rate would be half of that of output if $b = 1/3$. The output which is flattening out in 6.3 might continue to grow steadily if the labor input is increased. Th s is hardly contemplated in the case of the progress function where a given production outfit is presumably the background from which the data have been gathered. We have quite a different case here, requiring for effectiveness a change in equipment in order to realize the economies of scale or "increasing returns." This may also be regarded as a learning process, taking place in a longer time in consequence of the expansion of the market (Smith 1776) but with a recasting of productive apparatus. The decreasing cost in this case may also be described by expression 6.2, but the economic interpretation is now different from that of the learning function. It is rather intriguing that the value of $1-b$ as a scale effect, usually two-thirds, corresponds to the value of the learning b in the aircraft industry, one-third. Perhaps this is a coincidence.

The usual value of the scale effect, 2/3 (often explained by the proportion of surface to volume in a cylinder, the former related to cost and the latter to output) is quite widely used by engineers to estimate cost for different dimensions of the same equipment (especially in chemical plants).

Some economists (Young 1928; Kaldor 1972) have apparently tried to see the whole technical progress as a phenomenon of increasing returns (which, through external returns, affects the national output as a whole). If the scale effect is of the order of magnitude of the two-thirds formula, then this is unrealistic for the same reasons as were given in the case of the learning function. The growth of labor input would have had to be much greater than it was empirically in order to explain the observed growth of output and productivity. It might be argued that the scale effects are in reality greater than one-third; Verdoorn (1949) has calculated them from a regression of productivity on output, but this describes only the history of productivity and output and proves nothing.

Since the share of productivity in the growth of output is larger in historical experience than could be explained by either the learning function or the scale effects, there must be additional powerful elements of technical progress. We have not far to look for them; they are the innovations that involve major alterations in production methods and equipment. (They often involve an increase in scale, too, but that is not sufficient to include them simply under the heading of increasing returns.) We may, *cum grano salis* of course, say that the progress function describes a continuous process of small improvements that takes place without major changes in equipment or skill structure (disembodied technical progress); where as the innovations describe a discontinuous process that involves major changes in equipment and skill structure, perhaps even emergence of new industries or trades (embodied technical progress).

An innovation continues to act after its inception. It will spread from one firm to others, from one country to other countries. This is the process of diffusion. It may continue on a higher plane if new applications of the same novelty, for example in other industries, are found.

Diffusion by its very nature should tend to a limit and thus we feel intuitively that the effects of an innovation will die out after a time. There may be new influences from outside, however, from the inner development of science, or from the economy-society, which may set new tasks or provide new tools for observation. In this way we can understand that the process of growth in productivity or in variety of goods may be kept up on a more or less stable level.

Interaction of Technology, Science and Society

It is tempting to represent the discontinuous process of innovation by a graph as in network analysis. In this way we catch two essential features of the process: (1) *the ordering*. It proceeds by a succession of steps and each step can be taken only if certain other steps have been taken before. Each step represents a technique, a way of doing a thing, counting however, only the discontinuous innovations usually embodied in new equipment or at least major adaptations of equipment. (2) *the element of time,* which elapses between one step and the following one, which we have to regard as a random variable. There exists a critical path that determines the speed of the whole process.

The network analysis is, of course, a routine instrument in the planning of large R&D projects. What is suggested here, however, is rather its use as a formal tool in analyzing the history of technology. We may start from major innovations as ancestors and construct the trees of techniques that arose from them; they will interlock repeatedly because new techniques often arise from a combination of others. We may also start from a combination of today's techniques and trace their ancestry.

The use of oxygen in steel making may serve as illustration (See figure 6-1).

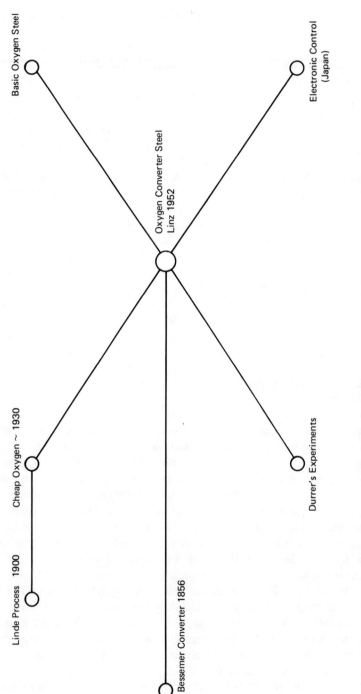

Figure 6-1. Succession of Steps in Technical Developments Leading to Oxygen-blown Converter Steel

The idea of using oxygen instead of air in the steel converter goes back to Bessemer himself (1856). It was impracticable as long as there were no methods of producing oxygen cheaply. This problem was solved in two steps, by the Linde process around 1900 and by its improvement about 1930. A water-cooled lance to blow the oxygen on the surface of the steel bath was used in experiments by the Swiss Durrer during the war. After the war this technique was perfected in Linz in 1949, and in 1952 commercial production started. In its course large-scale economies also came into play. The size of the converters was increased from 30 tons to 200 tons. The method was subsequently extended to the basic process (high phosphorous content in iron, formerly Thomas process). It was further developed by electronic control of the process in Japan. It favors the use of continuous casting and therefore in turn contributed to the development of this technique (Schenk 1974).

Figure 6-1 shows how a combination of several techniques is needed to produce a new result. We see also how an innovation produces further techniques in good time. There are three ways: (1) the extension to further fields (from the acid to the basic process); (2) the replacement of technique after a time by a completely new version of it (the Bessemer by the oxygen converter); (3) the creation of a new environment which stimulates further changes in related fields (continuous casting).

The graph of techniques might be made more complete if we include the whole development of science in it. This would bring some difficulties; though some scientific advances can almost be regarded as techniques (such as the thermionic valve), others are ideas, concepts, operations, theories. We can really only call them steps. It might be better to deal with science separately and to admit that there is frequent interaction between the two systems.

Even less could we include the economy in the graph of techniques, since its evolution could not be very well illustrated by a graph.

The techniques and their links have an economic dimension that is somewhere behind the graph: the amount of product produced by a technique as it diffuses in the course of time; the reduction of cost, especially of labor cost; the amount of equipment produced that embodies a technique; the increase in the rate of investment brought about by an innovation. The diffusion of a new technique through the economy determines its effect on (labor) productivity; it is also important for the learning effects—possible improvements of the techniques or adaptation to further uses.

The innovative process determines the growth of productivity and the possibility of increasing the standard of living. This is its effect on supply (it is equivalent to increases in the supply of factors such as labor). At the same time it creates new effective demand by stimulating investment. This, in fact, is essential for growth. Without outside influences such as war, armaments, foreign investment or innovations, the process of investment must run down owing to the self-defeating effects of accumulation on the utilization of equipment and the rate of profit. This is the picture given in Kalecki's (1954) theory of the trade cycle and the trend, and it offers a certain analogy to system theory. As Bertalanffy argues, a closed system comes to rest; growth requires a stimulus from outside. While he is thinking of an energetic system, the economic system as viewed here is a system of effective demand (under certain supply constraints).

The innovative process is thus the engine of long-term economic development; at the same time it is decisively shaped by the economy and by society. Bottlenecks in the economy may stimulate new techniques (innovations in spinning were stimulated by the demand for yarn due to the introduction of the flying shuttle in the industrial revolution). Society may set new tasks that direct the technological and scientific development, especially in an era in which organized R&D has become more and more important as compared with learning by doing (which it can never completely supersede, however). Most prominent are the tasks set by the science-conscious military, in view of the large support by public financing. A new kind of task comes from environmental policies (and conservation policies in the case of energy). This is probably only in its initial phase of development.

Society and the economy form the environment in which technology and science have to survive. This environment, however, to a large extent has been and is being created by the technical evolution itself. In consequence the distinction of ends and means is blurred. (Some authors have come to the same conclusion in their studies of policymaking, R&D planning, and so on; see Hirschmann and Lindblom 1962.) Sensitive people such as Samuel Butler or E.F. Schumacher (1973) have been deeply shocked by this.

All this leads us to the questions, Are there alternative techniques? Is there a choice of the path? The study of technological evolution cannot, logically, provide a straight answer to this, but it may provide a useful background for a judgment.

Classification in Economics

A stochastic process would seem to be a highly appropriate model of technical evolution, but it would have to be a very drastic simplification indeed. G.U. Yule (1924) has dealt with the parallel subject of evolution of species, using a linear birth process. He studied the number of genera containing 1, 2, 3, ... species, seeing that empirically this followed a Pareto distribution. His birth process, assuming that new genera were much more rarely produced than new species, explained the observed distribution.

Could anything similar be tried with techniques or combinations of them, that is, with industries and their subdivisions? There is surely a parallel here, because a main technique appears in various combinations and variations, just like a genus in various species. I seem to remember that Zipf worked on such a classification of industries, but the subject is fairly hopeless, since we have no adequate classification of industries or commodities or techniques. The classifications used for census purposes are distorted by the use of the decimal classification. I have often wondered why no economist or economic historian has ever tried to work out classifications like those for living organisms, related to the trees of evolution of techniques, and their embodiments in industries and commodities. As it is, our statistical knowledge of techniques is rather thin for the ambitious aim of a model.

The Diffusion of Know-How

It is characteristic of human techniques (know-how) that they can be passed on in principle to anybody anywhere. The pool available for recombination extends to all known techniques. However, the transfer is not without its difficulties. The know-how has to pass through the medium of society which is more or less permeable. It may be nearly impermeable.

The consequence is that diffusion is sometimes blocked, at other times suddenly released. After World War II Europe introduced in a relatively short time a great number of American techniques that had been in use there for some time already. Clearly the diffusion had been largely blocked before. In the interwar years there was comparatively little movement of goods and people; embodied technique was shut out for lack of exchange; nor was there much money in Europe for studying in the United States and there was no opportunity to work there. In the war years diffusion was completely blocked. Thus there accumulated a stock of know-how commonly available in the United States but not yet in continental Europe. After the war the conditions changed abruptly. Marshall aid and technical assistance initiated a flow of plant, machines, and expertise which subsequently continued in the climate of liberalization and growing foreign trade. European technique was lifted by quick steps to (nearly) the American level of technical efficiency in many fields. It was drawing on a stock at a rate at which it could not have produced innovations by current efforts (through R&D). This explains that productivity in manufacturing increased in European countries up to twice as much as in the United States in the fifties and sixties. Similar to Europe but even more extreme was the position of Japan. (Explanations similar to these are given by Maddison 1964; Gomulka 1971; Economic Survey of Europe 1969.) At least as important was another effect of the sudden diffusion. The adoption of the new technique necessitated or facilitated large industrial investments which accounted for a great part of the high volume of employment in Europe.

This wave by its nature had to subside after a time. There are still large gaps in know-how but not of the type that can easily be taken over. Also a new type of vehicle for technological transfer has in the meantime developed, the multinational concern, but this has hardly as much effect on investment as the process already described.

The general question of the medium of diffusion can only be touched here. The permeability depends partly on the amount and type of education. An elitist education with a small upper stratum of sophisticated scientists, and little between them and the rest of the population, will be highly disadvantageous. The permeability will be increased by a broad education with plenty of intermediary strata of different grades of education, which will serve as interpreters and carriers of information. A great strength of the United States is that is possesses this varied supply of intermediate strata of education. Again, a sharp division between science and engineering on one side, law and humanities on the other, makes the society less permeable.

The attitude toward R&D and the greater or lesser ease with which its results are turned to practical use are strongly affected by these factors.

References

Ayres, C.E. 1944,. *The theory of economic progress.* Chapel Hill: University of North Carolina Press.
Butler, Samuel. 1872, 1954. *Erewhon,* Harmondsworth: Penguin Books.
Economic Survey of Europe, 1969. Part I. New York, 1970.
Feyerabend, P. 1970. "Against method." *Minnesota studies in the philosophy of science*, vol. 4. Minneapolis: University of Minnesota Press.
Gomulka, St. 1971. Inventive activity, diffusion and the stages of economic growth. Aarhus, Denmark: Institute of Economics.
Hirschman, A.O., and Lindblom, C.E. 1969 "Economic development, research and development, policy making". In *Systems Thinking,* ed. F.E. Emery. Harmondsworth: Penquin books.
Kaldor, N. 1972. The irrelevance of equilibrium economics. *Economic Journal.* vol. 82. p. 1145.
Kalecki, M. 1954. *Theory of economic dynamics.* London: Allen and Unwin.
Kuhn, T.S. 1970. *The structure of scientific revolutions*. Chicago: University of Chicago Press.
Maddison, A. 1964. *Economic growth in the west.* London and New York: Twentieth Century Fund.
Schenk, W. 1974. "Continuous Casting of Steel." In *The Diffusion of New Industrial Processes,* ed. L. Nabseth and G.F. Ray. London: Cambridge University Press.
Smith, A. 1776, 1977. *An enquiry into the nature and causes of the wealth of nations.* Chicago: University of Chicago Press.

Schuhmacher, E.F. 1973. *Small is Beautiful*. London: Blond and Briggs.
Verdoorn, P.J. 1949. Fattori che regoleno lo sviluppo delle produttività del lavoro. *L'Industria*.
Young, A. 1928. Increasing returns. *Economic Journal*, vol. 38, pp. 527-542.
Yule, G.U. 1924. A mathematical theory of evolution. *Transactions of the Royal Society of London*, vol 33, pp. 21-87.

7 The Scale Factor in Research and Development

Mark F. Cantley and
Vladimir N. Glagolev

Levels of Scale

In many decisions one of the significant parameters may be a choice of scale. Particularly where major decisions are concerned, these scales arise infrequently, so that management is less able or likely to see them as familiar. One has to stand back and take a broad or long-term view to see groups of decisions, perhaps in widely different places and industries, as members of a common class. But such generalization may help to avoid repetitions of similar mistakes.

To start this process of generalization and categorization, certain levels of scale can be usefully distinguished.

Level 1: The engineering level or unit level.
a. The scale of a single unit of physical equipment.
b. The scale of a single product line (which might be produced by several separate units of equipment).

Level 2: The plant level. The scale of a single plant or factory (on, or based on, one site; but possibly containing several engineering units or product lines).

Level 3: The corporate level or organization level. The scale of a single organization.

Level 4: The national level. The scale of national economic programs and industrial complexes.

Levels 1 and 2 coincide in the case of a single-unit (or single-train) plant, which typically depends on a single major component.

Factors of Scale

The different "factors" indicate the many contexts within which any decision on scale has to be made.

1. Political. Possible conflict between desire for independence and need for economic efficiency.

2. Social. Implications for society and for working conditions of centralized or decentralized production.
3. Economic and technological. These are sometimes the only factors explicitly evaluated.
4. Organizational. Changes may be required following a change of scale in product or process.
5. Managerial. The nature of work and required skills may change.
6. Financial. The multiyear construction of large-scale plants or projects may require special financing arrangements.

Textbook and technical evaluations tend to stress only the economic and technological; but real decisions are often severely constrained by the other factors. For example, in determining the scale of a single plant, most of the *internal* factors tend to favor increasing scale, concentration, and specialization; but a contrary pressure may be exerted by external or environmental constraints, such as the availability of labor, the geographical location of material sources and output destinations, and the uncertainties that call for dispersed facilities to counter risks. Over recent years, the growth of urbanization and the economic potential promised by rationalization have tended to encourage a general growth of scale; but this has not been without its critics.

The Contexts of Scale-Related Decisions

On every level decisions on scale are taken in relation to an environment, whose size and characteristics should generally have a determining effect on the decision. Increases of scale taken with inadequate environmental consideration may lead to commercially unsuccessful products or projects, perhaps undertaken more for reasons of technology push than demand pull.

In the socialist, planned economies, the context of national, sectoral, and republic plans provides the targets and constraints within which the enterprise plans are prepared. Enterprise planning decisions are aimed at maximum efficiency and typically employ mathematical programming to calculate the best strategy for the location and size of productive facilities. Where economies of scale exist, the unit production cost is a function of the scale of production; and this nonlinearity makes the programming more complex but still feasible. Not all plants are necessarily of optimum scale in any general sense; considerations such as sources of raw materials and destination of outputs or constraints on locally available labor lead to choices of scale of plant related to local conditions.

In the market economies, scale decisions can have a significant relationship to competitive position; the pursuit of low unit cost by large-scale plant can lead to problems of excess capacity, underutilization of plant, falling prices, and cash flow problems. In oligopolistic markets behavior can become a complex game, in which all may lose unless the overall market is sufficiently price elastic. However, the stimulus of competitive conditions has been a driving force of technological innovation.

Whether in a planned or a market environment, planners still need an understanding of the relationship between scale alternatives and performance measures, including both productive efficiency and more indirect aspects such as flexibility.

Models and Generalizations

Many simple models of the economies of scale have been propagated. For example, many economics textbooks suppose the optimum size of firm or of plant to be the result of a balance between economies and diseconomies of scale, resulting in a U-shaped graph of efficiency against size. The diseconomies are usually less clearly identified than the economies. The latter are typically based on spreading fixed costs over a larger output, use of automatic and special purpose machinery, and similar physical or engineering factors. But the diseconomies are sometimes suggested as arising from administrative difficulty of maintaining appropriate controls and operating responsiveness. A simple and well-known example of a general model of economy of scale in process plant is that of the two-thirds power law. Gold is skeptical of such simple-minded formulas and of their widespread propagation in the literature of industrial economics. "Continuing reliance on convenient assumptions in place of exploring the realities of industrial practice has rendered the traditional theoretical approach to scale economics widely inapplicable in concept and all but trivial in its posited effects" [1, pp. 2-4].

Some authors, such as Bain [2] and Scherer [3], do in fact draw directly or indirectly on extensive empirical survey of industrial realities. Scherer draws useful distinctions between the different types of economy of scale in production that may be achieved—at single-product plant level, multiproduct plant level, and at multiplant level. He draws attention also to economies of scale in marketing, distribution, and other functions and finally to some of the pecuniary advantages accruing to large organizations. Often the organizations in an industry appear to be much larger and the pattern of ownership more concentrated than is indicated as necessary by considerations of production economies alone. But these studies tend to be *taxonomic* (naming the different categories of economies of scale) and *descriptive* (quantifying the degree of concentration of an industry, say, by the concentration ratio of the largest n plants to total industry), rather than *analytical* (explaining how the economies arise). The two-thirds power law does at least have a physical basis, albeit oversimplified, in relating capital costs to surface area and capacity to volume.

Gold suggests that "scale economies are derived from the increasing specialization of functions" and, hence, that *"scale be defined as the level of planned production capacity which has determined the extent to which specialization has been applied in the subdivision of the component tasks and facilities of a unified operation"* [pp. 4-5]. This is a strong and interesting proposal. According to Gold, it "raises doubts about the likelihood of finding scale effects which are universal among industrial processes covering the entire spectrum of physical and biological sciences, or over the entire size range of possible operating units within each" [p. 5]. Gold concludes: "It would appear that major new horizons must be explored before new advances in our understanding of the generalizable and nongeneralizable elements of changes in the scale of production are likely to be achieved" [p. 5].

While the views of those with detailed engineering expertise are rich in empirically based understanding, the technologist is not usually the best generalizer, for instance, in technological forecasting. His acquaintance with the trees can sometimes reduce his ability to view the woods. There is considerable evidence (reviewed by Sahal [4]) for the existence of progress functions, learning curves, or experience curves characteristic for each industry. These take forms such as "for every doubling in the unit cumulative total of items produced, there is a 20% reduction in unit cost." Such laws, if accurate, are very relevant to the dynamic aspects of problems of scale and have been so used by corporate strategists.

Some interesting evidence was assembled by Simmonds [5] in the context of various products in the chemical industry. His view is that the scale of plant rises inexorably in proportion to the scale of total output, although interrupted by "steps." Figure 7-1 is taken from his paper. It demonstrates not only the foregoing statement but the difficulty of a small country (Canada) in maintaining plants capable of coexisting with the larger plants of the adjacent larger market. This illustrates what has been said about the relationship of scale and context. The steps shown in figure 7-1 reflect the cycle referred to previously: expansion, overcapacity, price fall, market growth, full capacity utilization, expansion.

Within the market context the dynamics of the scale-growth process have been represented by Cantley and Glagolev [6] as in figure 7-2. This suggests a cyclic process whose components can be modeled by various theoretical models including those referred to before.

The Need for Research

Gold [1] has referred to the need for more empirical research. One industry in which there have been apparently well-documented economies of scale is that of the electricity-generating plant. Huettner [7], in particular, in a review of much

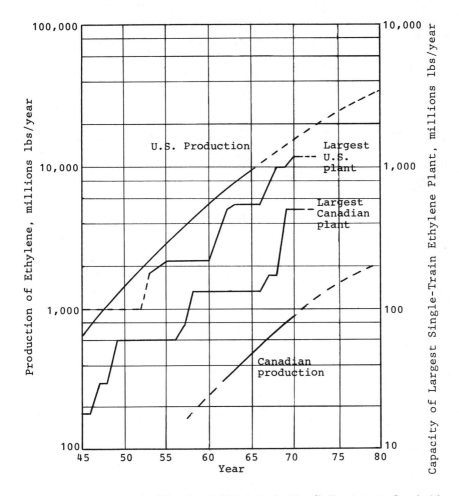

Source: W.H.C. Simmonds. "The Canada-U.S. Scale Problem." *Chemistry in Canada* 16 (October 1969), and W.H.C. Simmonds. "Stepwise Expansion and Profitability." *Chemistry in Canada* 16 (September 1969).

Figure 7-1. Growth of Market and Scale of Plant

previous work, confirmed the economies of scale in the fossil-fired plant but indicated that beyond 300 MW, the gains were relatively slight. Abdulkarim and Lucas [8] in the United Kingdom suggested that with the apparently lower reliability of large plants (which might be related to Gold's remark about scale and specialization—complexity), and with the extended construction periods and therefore greater forecasting uncertainty and planned reserve requirement, the stage might have been reached of actual *dis*economies of scale. This has been substantiated more recently and independently by Fisher [9] of General

148 Research, Development, and Technological Innovation

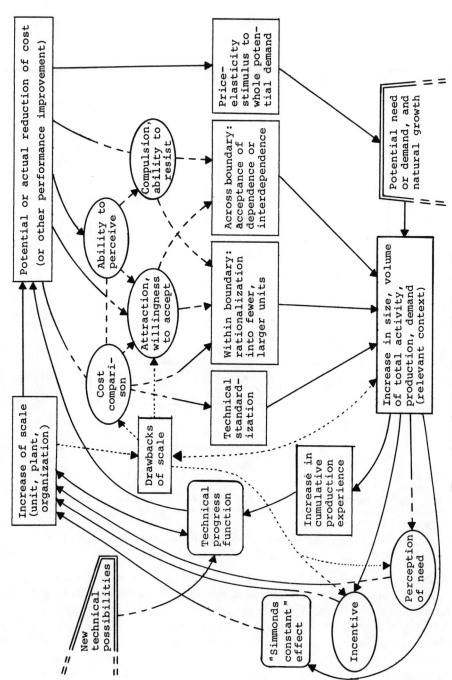

Figure 7-2. The Scale Growth "Mechanism"

Electric, using U.S. data. His work indicated an optimum as low as 175 MW (for fossil-fuel 2,400 psi 1,000F/1,000F plant), unless reliability of large plants improves or the construction period of large plants can be shortened.

Lee [10], also of General Electric, has suggested that the decline of reliability is an inevitable result of an accelerated pace of scaling up, in which components are replicated (giving greater opportunities for failure) and are stressed closer to their ultimate limits (giving increased probability of failure), while designs are not "matured" sufficiently before the next increase of scale is undertaken.

Fisher's work is intriguing in that it suggests a phenomenon of overshoot. The economies of scale in moving from 20 MW to 200 MW became so well-documented and established, not least in the minds of engineers and utility plant managers, that further increases of scale became uncritically encouraged, and larger-scale plants are ordered before the previous generation has been operationally evaluated for an adequate period. Bupp and Derian [11] suggest that the same phenomenon has been repeated with light-water nuclear reactors.

Nonetheless, opinion in the chemical industry until recently was that there are economies of production to be gained up to and beyond the largest plants yet built, the constraints on size still being only finance or market determined. This is a situation that International Institute of Applied Systems Analysis researchers are investigating in the context of ethylene plants.

Some indications of a changing perspective are indicated by Hansen of Shell (Germany) [12], with his (reported) remarks about that company's ethylene plants.

> Escalation of chemical plant construction costs has cut the advantage of large units to make lower-cost chemicals than small plants. F.H. Hansen, general manager of Deutsche Shell Chemie, speaking at a Society of Chemical Industry meeting in Taormina, Sicily, said that the cost/ton of installed capacity might triple by 1982 for a duplicate of a 450,000 tpy ethylene plant built in 1972. Deutsche Shell's experience at one location shows that the cost/ton of installed capacity declined markedly throughout the 1955-1968 period. This trend changed in 1972, when a 450,000 tpy unit started up. Its cost/ton of installed capacity was 50% greater than that of the 1968 plant. For the first time, a bigger plant did not make lower-cost ethylene. The cost/ton of ethylene produced was about 16% higher. There is no possibility of offsetting this cost by further economies of scale. Increases in the size of ethylene plants are technically feasible, but would lead to only marginal reductions in cost/ton. Technology improvements and large size, the two main factors that helped cut the cost of producing ethylene in the past, are likely to be less important than the costs of feedstock and building the plant. [p. 37]

This quotation emphasizes the economic and financial aspects. The technical aspects were emphasized by Friedman [13], on points strikingly similar to those made by Lee.

> Single-train continuous units for petrochemical and fuel products have been steadily increasing in size... [p. 80].
>
> However, evidence is beginning to be seen of size and dimensional limitations which will impose new demands on engineers and technology managers... [p. 80].
>
> Some recently reported mechanical failures have apparently been due to an inability to recognize that some previously insignificant design aspect had become significant when the size was increased... [p. 81].
>
> If the frequency of a specific atypical problem is proportional to single-train-plant capacity, more problems can be expected to be seen in tomorrow's plant as capacities continue to increase... [p. 83]

These quotations indicate some of the possible diseconomies of scale on economic and technical grounds, in certain major process industries. In summary, the following key topics deserve further analysis:

1. Capital and operating cost advantages
2. Flexibility and risk
3. Organizational structure, complexity and control
4. Sociological aspects of large scale organization
5. Learning curves, scale and technological innovation
6. Labour productivity

The implication of what has been said about electricity and ethylene is that R&D needs to consider the fifth topic as well as the first, in purely technical and economic terms. Moreover, the other four topics mentioned indicate the need to broaden the concept of R&D to embrace these statistical, organizational, control-theoretic, and sociological issues.

References

[1] Gold, B. Evaluating scale economics: the case of Japanese blast furnaces. *Journal of Industrial Economics* 23 (September 1974):1-18.

[2] Bain, J.S. Economies of scale, concentration, and the condition of entry in twenty manufacturing industries. *American Economic Review* 44 (March 1954):15-39.

[3] Scherer, W. *Industrial market structure and economic performance.* Chicago: Rand McNally and Co., 1971.

[4] Sahal, D. A theory of progress functions. *Transactions of the American Institute of Industrial Engineers.* 1979.

[5] Simmonds, W.H.C. The Canada-U.S. scale problem. *Chemistry in Canada* 16 (October 1969):39-41.
[6] Cantley, M.F., and Glagolev, V.N. Problems of scale: the case for IIASA research. International Institute of Applied Systems Analysis RM-78-47, 1978.
[7] Huettner, D.A. Scale, costs, and environmental pressures. In B. Gold, ed., *Technological change: economics, management, and environment.* Oxford: Pergamon Press, 1975.
[8] Abdulkarim, A.J., and Lucas, N.J.D. Economies of scale in electricity generation in the United Kingdom. *Energy Research* 1 (1977):223-231.
[9] Fisher, J.C. Size-dependent performance of subcritical fossil generating units. Electric Power Research Institute, October 1978; and amplification of this in public lecture at International Institute of Applied Systems Analysis, 10 October 1978.
[10] Lee, T. Public lecture at International Institute of Applied Systems Analysis, 5 December 1978.
[11] Bupp, I.C., and Derian, C. *Light water: How the Nuclear Dream Dissolved.* New York: Basic Books, 1978.
[12] Hansen, F.G. As reported in *Chemical Week,* 17 December 1975. Reprinted by special permission. Copyright 1975 by McGraw-Hill, Inc., New York, N.Y.
[13] Friedman, G. Large petrochemical plants are using tomorrow's concepts, *Oil and Gas Journal,* 12 December 1977. Reprinted with permission.

8 Diffusion, Innovation, and Market Structure

Stephen Davies

This chapter addresses the long-running debate within industrial economics concerning the influence of firm size and concentration on the technical progressiveness of industry. In particular, I shall be concerned with two aspects of technical progress: the *diffusion* and (to a lesser extent) the *innovation* of new processes.

The main stimulus for this study is a recently completed study by Davies (1979) of the diffusion of new processes in U.K. manufacturing industries. Parts of the theoretical and empirical analysis of that study are of direct relevance to the following two questions. First, within industries, do larger firms tend to adopt or imitate new processes more or less rapidly than small firms? Second, does the level of concentration affect the speed at which the industry diffuses new processes? In both cases I shall explore and contrast the relevant findings of earlier diffusion studies. In so doing, I hope to provide a survey to complement those already in existence on market structure and other elements of technical progress.[1]

For "innovators," firms who instigate diffusion by first introducing new processes into their industries,[2] I investigate data initially collected as a by-product (and therefore not previously analyzed) of my earlier diffusion study. In the light of these data, two further questions are considered. Do large firms perform their share of innovating? Is their performance in this context affected by the level of concentration in their industries?

Diffusion

Firm Size and the Speed of Adoption

Following the introduction of a new process by the innovator, what factors determine how rapidly other firms in the same industry follow suit and adopt that process? The first known large-scale empirical investigation of this question was provided by Mansfield (1963a) who collected data for 167 firms adopting fourteen different processes in various U.S. industries.

Define as t_{ij} the numbers of years that the jth firm waits (after the initial introduction by the innovator) before adopting new process i. Then Mansfield suggests six characteristics of the firm or process that might determine t_{ij}.

153

Abstracting from four of these variables, which were subsequently found to be statistically insignificant, one can write his model as

$$t_{ij} = a_{0i} \, S_j^{a_1} \, H_{ij}^{a_{2i}} \, e_{ij} \, , \tag{8.1}$$

where H_{ij} is the profitability of process i for firm j and S_j is the size of firm j, e_{ij} is an error term and a_{0i}, a_1, and a_{2i} are parameters.

Because of data deficiencies he was unable to observe H_{ij} for all firms and processes, and for those processes where data were available, a_{2i} was only significantly different from zero in two cases. For both these processes, however, the estimate a_{2i}, was consistent with the hypothesis that firms for whom the new process was most profitable adopted it more rapidly.

Here, however, we are more concerned with the S_j variable. This was found to be a consistently significant *negative* determinant of t_{ij}, regardless of which other explanatory variables were included in the equation. In Mansfield's preferred form of the equation, $\hat{a}_1$ takes the value of -0.4, implying that large firms do tend to adopt new processes more quickly than small firms.

This type of equation has been reestimated repeatedly in subsequent research, for other samples of processes, industries, and countries; other researchers have nearly always confirmed Mansfield's findings.[3] Firm size is usually found to be a significantly negative determinant of t_{ij} (larger firms adopt more quickly), but there has been little success at identifying other causes of interfirm differences in the speed at which a new process is introduced.

Mansfield suggests three reasons for this observed relationship between S_{ij} and t_{ij}. (1) Because of their greater size, there is a higher probability that larger firms will need to replace old equipment at any point in time. (2) Again because of their greater size, larger firms are likely to encompass a wider range of operating conditions than smaller firms. As some new processes have only limited applicability initially, there is more likelihood, therefore, that large firms will have the appropriate operating conditions for adoption of the process in its earlier years. Both these reasons suggest earlier adoption for large firms merely because of greater opportunities. In addition, however, (3) the greater financial and technical resources of large firms may better enable them to bear the costs and risks of early adoption.

My contribution (Davies 1979, chap. 6.) to this area of the diffusion debate produces similar results concerning the influence of firm size, although it is based on a different theoretical framework, using an explicit theory of decision making at the firm level. While I accept Mansfield's discussions of the benefits of large scale, I suggest that there is an even more compelling reason why large firms should tend to adopt more rapidly. An examination of the technical literature and discussions with the suppliers of twenty-two new processes in the United Kingdom suggested that most, if not all, of these processes exhibited some sort of scale economies effect (Davies 1979, chap. 3). In the case of large lumpy processes, such as furnaces and kilns, this effect derived from economics

Diffusion, Innovation, and Market Structure

in capital and operating costs as the scale of adoption increased. In other cases, such as looms, there exist certain economies in the adoption of large numbers of the processes. In yet other cases (such as supplementary devices to speed up paper machines), returns from adoption appear to be greater for larger and more specialized existing processes (such as existing paper machines). Thus assuming that larger firms tend to employ larger and more units of capital equipment, one can give technical reasons why they should derive greater returns from most new processes.

The theoretical model reflects this scale economy effect and other size-related effects and generates the prediction described in figure 8-1. More precisely, it predicts that at any point in time within a given industry the probability that a firm will have adopted a given new process is a monotically rising S-shaped function of firm size.[4] This prediction is found to be consistent with data collected on the diffusion of the sample of processes mentioned. (For each process, for at least one point in the diffusion period, all potentially adopting firms were grouped into various size classes. In all cases it was found that the larger the mean size of the class, the higher the proportion within size classes having adopted by that date. Moreover, the functional form of the relationship between proportions and size was consistent with the S-shape shown in figure 8-1.)

With little doubt, then, virtually all known empirical and theoretical work in this area points to earlier adoption by large firms. However, this need not necessarily indicate that large firms are more dynamic or progressive in this context. After all, scale economies and Mansfield's arguments suggest built-in advantage for large firms. In these circumstances, I am inclined to agree with

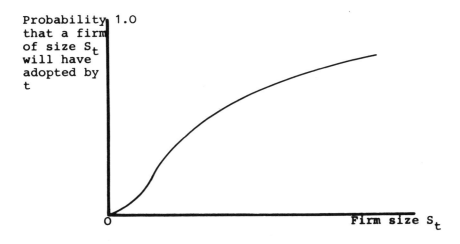

Figure 8-1. Prediction Generated by the Theoretical Model

Weiss, who, when discussing Mansfield's results, suggests that "the more basic question is whether the larger firms were quicker to imitate than groups of smaller firms with about the same number of investment decisions to make" (1971, p. 396). Unfortunately there is no way to employ Mansfield's and others' estimates of a_1 to provide an answer to this question. On the other hand, such is the nature of the relationship in my model (described by figure 8-1) that it is a relatively easy algebraic exercise to estimate the probability that a firm of some size X will adopt a new process more rapidly than both of two smaller firms, each of size $X/2$. Using estimates of the slope parameters of the curve in figure 8-1 for each of the sample processes, this probability is less than 0.5 in all but two of the sample of twenty-two processes. In other words, for nearly all the sample processes, a firm of any size will be slower on average than at least one of two firms of equal size, each exactly half the size of the first firm. Interestingly, this result is general, in that it does not depend on the value of X.

On the basis of this simple test, we should be somewhat wary of arguing that the diffusion evidence is necessarily consistent with greater progressiveness on the part of larger firms.

Industrial Concentration and Speed of Diffusion

Let us now turn from the performance of individual firms within a given industry to differences between industries in the speed with which they diffuse new processes. This leads directly into an empirical construction common to many past studies: the aggregate diffusion curve. This curve is usually employed to describe the growth over time in the proportion of firms within an industry having adopted (or imitated) a given new process; the time origin of the curve is defined by the date of first introduction by the innovator.

For nearly all processes the diffusion curve is found, empirically, to follow an S shape. This has led researchers to fit standard S-shaped time trends to the data, while identifying the slope parameters of those curves as measures of the speed of diffusion.

The most common of these standard trend curves is the *logistic*, first suggested by Griliches (1957) and then adopted by Mansfield (1961) and many others. The equation of this curve is

$$m_t/n = \left\{ 1 + \exp(-\alpha - \beta t) \right\}^{-1} \tag{8.2}$$

where m_t is the number of firms having adopted a given process at time t, of an industry population of n firms. α and β are the parameters of the curve, with β in particular determining the slope of the curve. As can be seen from figure 8-2a, larger β values imply steeper diffusion curves and thus more rapid diffusion. This has led to the convention of defining β as the speed of diffusion.[5]

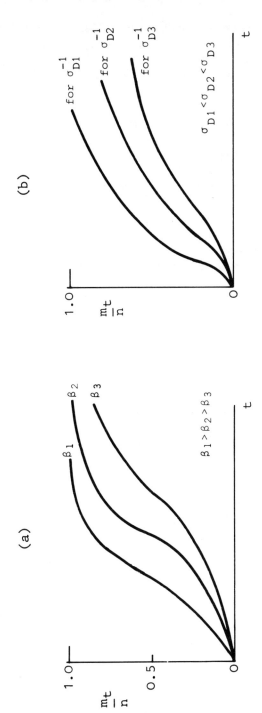

Figure 8-2. Diffusion. (a) Logistic, or Cumulative Normal. (b) Cumulative Lognormal.

However, as an alternative, I have suggested (1979, chap. 4.) that the diffusion curve follows one or another of two other S shapes, depending on the nature of the process concerned. Thus for a type of process labeled group B, I argue that diffusion grows along a *cumulative normal* time path, which can be written as

$$m_t/n = N(t \mid \mu_D, \sigma_D^2) \quad \text{for group B processes} \tag{8.3}$$

In other words, as time proceeds, diffusion describes a cumulated normal curve with mean μ_D and variance $\sigma^2{}_D$.

As it happens, this curve, like the logistic, is symmetrical, and although there are slight differences between the tails of the two curves, here we shall assume a rough equivalence of the two. Thus figure 8-2a may also be interpreted as referring to a family of cumulative normal curves. In this case it is the magnitude of σ_D^2 that determines the slope of the curve; the steeper curves refer to smaller σ_D^2 values. Thus we shall designate $1/\sigma_D$ as a measure of diffusion speed in this case.

Now a group B process is designated as one involving relatively large capital outlays by the adopter and characterized by fairly complex new technology. On the other hand, for relatively inexpensive and uncomplicated processes (often of a supplementary nature), it is suggested that diffusion follows a positively skewed *cumulative lognormal* time path. This group A diffusion curve can be written as

$$m_t/n = \Lambda(t \mid \mu_D, \sigma_D^2) \quad \text{for group A processes,} \tag{8.4}$$

where $\Lambda(\cdot \mid \mu_D, \sigma_D^2)$ denotes the lognormal distribution function, mean μ_D^2, variance σ_D^2. This means that the diffusion curve is cumulative normal if plotted against $\log t$ or as portrayed in figure 8-2b, if plotted against t. Again $1/\sigma_D$ is interpreted as a measure of diffusion speed, and figure 8-2b shows a family of these curves. The steeper curves refer to lower magnitudes of σ_D^2.

Both predictions 8.3 and 8.4 derive from the development of the theoretical model mentioned earlier. Basically, the alternative predictions arise mainly from a hypothesized difference in the nature of the learning curve between the two types of processes.

As can be seen, the essential difference between 8.4 on the one hand and 8.2 and 8.3 on the other is that the lognormal curve typically implies an earlier and more rapid takeoff in diffusion but, thereafter, a much more noticeable slowing down in the growth of diffusion. Thus, while many firms will adopt the typical group A process fairly quickly after its innovation, many years later a large number of other firms may still not have adopted the process, and some will probably never adopt it.

In the diffusion literature, there are a number of case studies in which the logistic curve has been fitted to data on the diffusion of different, usually major,

Diffusion, Innovation, and Market Structure 159

processes (group B using my terminology).[6] Judged simply on the basis of $\bar{R}^2$, the logistic usually provides a satisfactory fit. However, only in a few cases have sufficient processes and industries been examined to enable the researcher to investigate the causes for inter-industry-process differences in diffusion speed (as represented by estimates of the β parameters).

Mansfield's own seminal contribution (1961) included a sample of twelve processes in four industries. Using regression analysis of the twelve β estimates, he was able to show that diffusion speed is directly related to the profitability and inversely related to the cost of the process, but he had insufficient observations (only four) to test adequately hypotheses concerning the influence on β of the characteristics of the adopting industry. Nevertheless, he did feel able to suggest that his estimated β values were consistent (although not quite significantly so) with a positive relation between diffusion speed and the "extent of competition" in the four industries (after allowing for differences in the profitability and cost of the processes). Moreover, in a later case study of another process in a different industry (Mansfield et al. 1971), he found that its diffusion speed, relative to those of the original sample processes, was also consistent with this hypothesis.

A more recent study by Romeo (1977), following Mansfield's methodology exactly, has provided some more definitive evidence. Romeo's data related to the diffusion of a single process but in ten different sectors of the U.S. engineering industry. Having first fitted the logistic to these data, regression analysis of the ten β estimates identified three significant explanatory variables. The most interesting of these for present purposes is σ_s^2, the variance of log firm size within the adopting sector, which was found to be an inverse determinant of β. Since we can interpret σ_s^2 as a measure of concentration (more correctly, σ_s^2 is a measure of size inequalities, which is one dimension of concentration; the other is firm numbers) this is consistent with an inverse relationship between diffusion speed and concentration.[7]

My alternative predictions 8.3 and 8.4, have been fitted to data on the diffusion of the twenty-two processes (in thirteen industries) mentioned in the previous section. (Unlike most previous studies this sample not only comprises major new processes but also includes eight of the group A type). Both curves perform well for the relevant processes (in terms of estimated $\bar{R}^2$ and Durbin-Watson (D-W) statistics). On the other hand, as might be expected, neither curve provides a satisfactory fit when applied to inappropriate processes (such as the cumulative normal for group A processes). Interestingly, employing standard test statistics, the logistic appears to be a misspecification of functional form when fitted to diffusion data on all the group A processes.

Further cross-sectional regression analysis of the estimated slope parameters for this sample identifies four main (significant) determinants of diffusion speed. One of these corresponds to Mansfield's profitability variable, but two others are of more direct relevance here. Diffusion tends to be more rapid (σ_D is smaller) when there are fewer firms in the adopting industry (since diffusion is measured as a proportional variable, this is not a trivial finding) and when the size inequalities between firms are smaller (where size inequalities are measured by the variance of log firm size).

At this point I offer only two brief comments on these results. First, the significant inverse role for size inequalities confirms Romeo's finding, although the rationalization differs. Second, the additional inverse influence of firm numbers suggests a complex relation between diffusion speed and concentration (itself related positively to inequalities but inversely to numbers).

Concentration and Diffusion Speed:
The Theory

Having established the empirical "fact" of a relationship (of some sort) between speed of diffusion and concentration, we now examine the theoretical explanations offered for these results. But to do this we must examine the theoretical underpinnings of the alternative time curves 8.2-8.4.

Consider the logistic curve. The most elegant mathematical explanation as to why diffusion should follow logistic growth is due to Mansfield (1961). If we simplify his model (which, incidentally is independent of his explanation of interfirm differences, equation 8.1), we may identify the following crucial hypothesis:

$$(m_{t+1} - m_t) / (n - m_t) = \beta m_t / n, \quad \beta > o, \tag{8.5}$$

where m_t and n are as defined earlier.

In words, this hypothesis is that within a short time period, t to $t+1$, the proportion of nonadopters deciding to adopt a given process (already in use by at least one other firm in the industry) will be proportionate to the proportion of firms already using the process (the level of diffusion) at time t.

Mansfield rationalizes this equation (which is identical to that employed in simple epidemic models of contagious diseases) by implicitly discussing the reasons why nonadopters should revise their attitude toward the new process. More specifically, he argues that the proportion of nonadopters adopting is greater when m_t/n is higher, for three reasons. "As more information and experience accumulate, it becomes less of a risk to begin using [the new process]. Competitive pressures mount and "bandwagon effects occur." (1968, p. 137). He goes on to develop the informational point by suggesting that "where the profitability of using the innovation is very difficult to estimate, the mere fact that a large proportion of its competitors have introduced it may prompt a firm to consider it more favourably" (1968, pp. 137-138).

With equation 8.5 it is a trivial step to the logistic growth curve; t to $t+1$ is assumed to be very short and thus discrete time is replaced by continuous time. The resulting differential equation then has the logistic curve 8.2 as its solution.

Consequently, Mansfield's discussions of the likely determinants of the magnitude of β in 8.5 amount to hypotheses concerning the determinants of diffusion speed.

In fact, his model points to a number of possible determinants of β, including the profitability and cost of the process and, most important in the present context, the extent of competition within the adopting industry. He suggests that β should be higher, all else equal, in "markets which are more keenly competitive" (1968, p. 138). This hypothesis presumably derives from the "competitive pressures" argument. For instance, if one assumes that the process is cost saving, the cost disadvantage of nonadopters may well not result in any loss of market share in industries characterized by price fixing (collusion, price leadership, or the kinked demand curve). In those circumstances the incentive to adopt is not necessarily lower than in more competitive industries, but the pressures on nonadopters may be substantially smaller. Likewise, in industries characterized by highly differentiated products, price reductions on the part of adopters, passing cost savings on to the consumer, will make smaller inroads into the market shares of nonadopters.

At any event, both Mansfield's and Romeo's empirical findings on the influence of competition (or concentration) on β appear consistent with this hypothesis.

Having said this, one might also develop an opposing hypothesis within the framework of this model, in terms of the diffusion of information. For instance, in industries comprising only a few firms, and perhaps practicing some form of implicit collusion, there may exist the ideal conditions for the rapid interchange of reliable information concerning the characteristics of new processes. At the least, the task of the supplier of the process in advertising his new process (perhaps mainly through the employment of a small number of technically qualified salesmen) is considerably eased in such cases. Moreover, while there is no evidence of such an effect predominating in either Mansfield's or Romeo's results, it is certainly consistent with my finding concerning the inverse relation between diffusion speed and firm numbers.

Let us consider now the model underlying equations 8.3 and 8.4. As we shall see, this too suggests a potential influence for concentration through competitive pressures and information diffusion. In addition, however, a third, more subtle effect, which results from the prediction concerning the role of firm size, may be identified.

Fortunately, the relevance of this alternative in the present context can be described by employing what amounts to a very special simple case, described by the following assumptions.

1. Assume that a potential adopter will introduce the new process (already in use elsewhere in his industry) only when his expectations of the return from adoption R^1 is equal to or exceeds a critical or target rate of return R'', which he uses to assess risky new investments. In that case, i will have adopted by t only if[8]

$$R_{it}^1 > R_{it}''. \tag{8.6}$$

2. For reasons discussed earlier, assume that the larger firm i is, the higher are expected returns. Indeed here we shall assume that R_{it}^1 depends only on i's size according to the simple relationship

$$R_{it}^1 = a_t S_t, a_t > 0. \tag{8.7}$$

(In the full form of the model the possibility of a nonlinear relationship is permitted. Moreover, firm size is specified as only one of many potential determinants of R_t^1. Others would include the precise nature of the products, existing processes, inputs, and management of i, which will influence his actual rate of return, and the ability of i to collect and interpret information about the new process, which will determine how far the expected rate diverges from the actual.)

3. R_t^1 is assumed to be higher the later the vintage of the process adopted; thus a_t increases monotonically with time. This is partly due to improvements over time in information (as implied in the Mansfield model), arising from contacts with other firms already using the new process and with the firm(s) selling the new process. But probably more important is the phenomenon of learning by doing, mainly on the behalf of the suppliers of the process.

However, it is argued that the suppliers' learning curve will depend on the nature of the process. Thus for fairly technologically simple and inexpensive processes built off site, improvements in specification from learning by repetition will be initially substantial, but after this initial spurt of postinvention improvements, the technology will become fairly stable at an early stage in the diffusion period. (This is the group A process; examples in my sample include devices or machines designed to increase the speed of paper machines and various stages in the weaving process). On the other hand, the more expensive processes, based on more sophisticated technology and produced on a one-off basis, often necessitate lengthy periods of installation on the adopter's site. Usually, adopters will only require one or two units of these lumpy innovations, and major modifications are often necessary to meet the requirements of different adopters. From the manufacturers' point of view this means considerable heterogeneity in their installations. Consequently, learning from repetition is more limited in the early years, and more often than not teething troubles are encountered. Thereafter, however, as the manufacturer begins to increase his knowledge about his customers' operating conditions, a period of rapid and sustained learning often takes place. This is identified as the group B learning curve. (Examples of group B processes in my sample were the oxygen steel-making process and tunnel kilns).

For reasons of brevity, we shall consider here only the B type of process. The preceding discussion might suggest the following extension of 8.7:

Diffusion, Innovation, and Market Structure

$$R_{it}^1 = a_0 e^{g^1 t} S_{it}, \quad a_0 > 0, g^1 > 0. \tag{8.8}$$

4. Let all firms in the industry employ the same target rate in assessing adoption. (This stark assumption is not required in the full form of the model.) But, following Mansfield's arguments, we assume that because of mounting competitive pressures and reductions in subjective risk assessments, all non-adopters relax this rate over time:

$$R_{it}'' = \overline{R}_0'' e^{-g'' t}, \quad \overline{R}_0'' > 0, g'' > 0. \tag{8.9}$$

Now following substitution of 8.8 and 8.9 into 8.6 and then rearranging, we find that firm i will have adopted by time t only if it is of size not less than some critical level s_{ct}; that is,

$$i \text{ has adopted by } t \text{ if } S_{it} > S_{ct} \tag{8.10}$$

$$\text{where } S_{ct} = ae^{-gt} \text{ and } a = \overline{R}_0''/a_0', \ g = g^1 + g''. \tag{8.11}$$

Equation 8.10 allows us to move immediately to a prediction concerning the level of diffusion at t. Since only those firms of size not less than s_{ct} will have adopted, diffusion is given by the proportion of the firm size distribution to the right of s_{ct} (see figure 8-3). However, quite clearly this proportion will increase over time, given the form of s_{ct} in 8.11.

We can formalize and develop this prediction by invoking the fairly common assumption of a lognormal firm size distribution; that is, let $\log S_t$ be normally distributed with mean μ_s and variance σ_s^2 (both assumed here to be fixed over time). In this case diffusion at t is given by the proportion of that normal distribution to the right of $\log S_{ct}$. This we shall write as

$$m_t/n = 1 - N(\log S_{ct} \mid \mu_s, \sigma_s^2). \tag{8.12}$$

In order to examine the time path of diffusion, we substitute in the earlier expression for S_{ct} from 8.11 and express in terms of the standard normal:

$$m_t/n = 1 - N\left\{\frac{\log a - gt - \mu_s}{\sigma_s} \mid 0, 1\right\} = N\left\{\frac{\mu_s - \log a + gt}{\sigma_s} \mid 0, 1\right\}. \tag{8.13}$$

(The final expression in 8.13 is derived from the symmetry of the standard normal.) Next, we note that 8.13 may be expressed equivalently as

$$m_t/n = N(t \mid (\log a - \mu_s), (\sigma_s/g)^2). \tag{8.14}$$

In words, then, diffusion at time t is given by the area under the normal curve with mean $(\log a - \mu_s)$ and variance $(\sigma_s/g)^2$ to the left of t. Therefore, as t

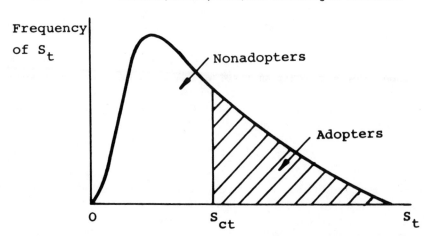

Figure 8-3. The Level of Diffusion at t and the Size Distribution of Firms

increases, this area increases and m_t/n sketches out a curve that is the cumulated form of the normal distribution.

This is, of course, the cumulative normal growth curve as described earlier for a group B process (see equation 8.3). Since we have already explained that the variance of this function is an inverse measure of the slope of the curve, the variance in 8.14 (equivalent to σ_D^2 in 8.3 tells us that diffusion will be faster (a) the larger is g and (b) the smaller is σ_s^2.[9]

Consider the implications for this investigation. First, the g parameter suggests a similar role to that in the logistic model for the effects of concentration through competitive pressures and information flows. Here these effects, along with postinvention improvements in the technology are reflected by the magnitudes of g^1 and g''. Second, the σ_s^2 parameter suggests an additional role for concentration which is absent in the epidemic model. Its presence in the expression for diffusion speed in this model suggests that, all else equal, diffusion will be slower the greater are size inequalities (and thus concentration) in the adopting industry. This prediction is certainly borne out by the results cited earlier (Davies 1979; Romeo 1977).

This additional role for concentration, while not intuitively obvious, follows logically from the earlier findings that firm size is an important determinant of the speed at which individual firms adopt a given process. Quite clearly, the diffusion period is longer the greater the differences between firms in the speed at which they adopt. In turn, these differences will be greater the greater are the differences between firms in their size (the larger is σ_s^2).

Further, my finding that firm numbers is an inverse determinant of diffusion speed can be explained in this model by an influence on g^1 (and accepting the earlier expressed hypothesis that information diffusion and exchange may be more difficult in high-number, unconcentrated industries.

Finally, by way of a postscript, I might briefly mention that the full form of the model uses a set of less restrictive assumptions. For instance, in practice there are many determinants of R^1 in addition to size; similarly there are interfirm variations in the target investment rate R_t'' (some of which may in turn be related to size). These modifications lead to interfirm variations in critical size S_{ct} and, along with equation 8.10, form the rationale for the prediction shown earlier in figure 8.1. Similarly, the parameters of the firm size distribution are allowed to vary, over time in response to growth in the adopting industry. In addition an alternative form of learning curve is employed for group A processes, and for all processes, both R_t'' and R_t^1 may vary across the business cycle.

These modifications (described in more detail in Davies 1979) generate additional predictions concerning the determinants of diffusion speed and the shape of the diffusion curve. Crucially for present purposes, however, the predictions of the simple form concerning the influence of concentration on the speed of diffusion continue to hold in this more developed model.

Innovation

Given then that there are theoretical arguments and empirical evidence that firm size and concentration have some influence on the diffusion of new processes, it is natural to ask whether similar relationships exist at the innovation stage.

Most previous research in this area has employed surrogate measures for innovative activity by firms or industries (patents, R&D expenditures), and there are still remarkably few studies based directly on data relating to numbers of processes or products innovated. One such study, however, of monumental proportions, is Mansfield's (1963b) investigation of the firms responsible for innovating approximately 150 major new processes or products in three U.S. industries in two time periods: 1919-1938 and 1939-1950. Among the findings of this work, we note here only two. (1) The largest four firms in the coal and petrol industries were responsible for a larger share of the respective industry's innovating than of its productive capacity. On the other hand, the four largest steel producers were responsible for fewer. In a later study (1971), Mansfield found, further, that the market share of the four largest pharmaceutical firms exceeded their share of the industry's innovations. (2) Maximum innovative activity occurred at approximately the size of the sixth largest firm in the petrol and coal industries but at a very small scale in the steel industry. For the pharmaceutical industry, innovative activity peaked at the tenth largest firm in 1935-1949 and at smaller scales in 1950-1962. We can add a third finding of Williamson's (1965) who, using Mansfield's data, found that the ratio of the top four firms' share of innovations to their share of capacity declined as concentration increased.

These results continue to dominate most recent literature surveys on market structure and innovation (for instance Kennedy and Thirlwall 1972, Weiss 1971, and Kamien and Schwartz 1975). Therefore, as a small contribution to this particular debate I shall provide a brief analysis of some relevant data collected in connection with (but unreported in) the diffusion study by Davies (1979). Table 8-1 reports, for each of eighteen of the processes considered in that work, the size of firm responsible for innovation (the first adopter of each process).

Table 8-1
Relative Sizes of Innovators

Process	Size of Innovator	Mean Firm Size in Industry	Innovator's Size Class[a]	C_j[a]	NS_j[a]
Numerical control[b]	6,500	614	Large	54	13
Photoelectrically controlled cutting	6,653	1,048	Large	19	6
Shuttleless looms[d]	750	168	Medium	34	4
Numerical control[e]	5,000	1,686	Medium	59	13
Gibberellic acid[f]	110	45	Medium	46	1
Numerical control[g]	1,200	527	Medium	56	13
Basic oxygen process[h,i]	8,400	8,628	Small	59	4
Wet suction boxes[j]	700	768	Small	53	2
Vacuum melting[h,k]	1,100	1,336	Small	70	8
Foils[j]	429	768	Small	53	3
Synthetic fabrics[j]	429	768	Small	53	1
Computer process control[h,j]	323	768	Small	53	4
Special presses[j]	212	768	Small	53	4
Computer typesetting[l]	325	462	Small	25	4
Tunnel kiln[h,m]	27	98	Small	45	8
Vacuum degassing[h,i]	1,500	5,714	Small	69	8
Automatic track lines[h,n]	10,400	27,536	Small	90	19
Continuous casting[h,i]	350	7,609	Small	56	9

Source: S.Davies. *The Diffusion of Process Innovations*. Cambridge: Cambridge University Press, 1979.
Note: Relative size is measured by employees at date of study.
[a]See text for definitions of these terms and symbols.
[b]Manufacture of turning machines
[c]Shipbuilding
[d]Textiles weaving
[e]Turbine manufacture
[f]Malting
[g]Printing press manufacture
[h]Steel
[i]Group B process
[j]Paper and board
[k]Special Steels
[l]Local newspapers
[m]Bricks
[n]Car manufacture

The conclusions to be drawn from such a small sample are nothing more than suggestive, but a number of interesting findings do emerge from simple statistical analysis.

Consider first the following simple hypothesis. *In each industry, each firm has a propensity to innovate which is proportional to its share of its industry's employment* (used here as the measure of size in the absence of widespread reliable capacity or sales data). As a test of this hypothesis, in each industry, firms have been ranked in order of size and then categorized into three groups: large, medium, and small. Each group accounts for one-third of total industry employment. Then, for each process, column 4 in the table reports within which group (in the relevant industry) the innovator is located. Clearly, this hypothesis leads to the expectation that innovators will be spread evenly over the three groups.

As can be seen, however, these expectations are hardly confirmed. Only two of the processes were innovated by large firms, four by medium, and, perhaps surprisingly, twelve by small firms. As it happens we have (just) sufficient observations to make a chi-square significance test of this result worthwhile and indeed the chi-square value is high enough (9.7) to allow us to reject the hypothesis at the 5 percent level; that is, on the basis of these figures, large firms have not performed their share of innovating.

Second, consider the identity of the processes innovated by small firms. The sample as a whole contains seven group B processes; of these, all were innovated by small firms. Recalling our earlier definition of a group B process as one involving a large capital outlay and fairly sophisticated new technology, this is a truly surprising result. It is presumably the group B process for which large-scale firms are best suited as innovators (following conventional arguments about their greater technical know-how and ability to finance risky and high-cost new projects). This said, however, there is a precedent for this result in the existing literature. Adams and Dirlam, in their famous case study (1966), of the basic oxygen steel process in the United States, find that "the invention was neither sponsored nor supported by large, dominant firms" and "it was a small firm that first innovated the new process in the U.S." Their explanation of this finding deserves some attention, given our result. They suggest that "it may well be that the structural and behavioural characteristics of oligopolized industries *prevent* the dominant firms from pioneering. Instead, the small firms may be the innovators because, unlike their giant rivals, what they do in the way of cost reductions is unlikely to cause so violent [a] disturbance of the status quo" (p. 188).

If this hypothesis is accepted and coupled with our earlier findings on firm size, an interesting picture emerges. While larger firms tend to be quicker to adopt processes already introduced by other firms in their industry, fears of upsetting the status quo typically dissuade them from "making the first move" and taking the role of innovator.

Finally, three simple hypotheses have been investigated using regression analysis of the data of table 8-1. Specifically, is there a tendency for the innovator to be relatively smaller (1) in more concentrated industries, (2) for less profitable processes, and (3) where competition between suppliers of the new process is limited? The first hypothesis needs little additional discussion. Clearly, it is suggested by Williamson's results and Adams and Dirlam's argument. The second might be true if larger firms are only prepared to risk innovation given high potential returns. The third is based on the following argument. When the new process is supplied by a monopolist, the latter might be prepared to ease his new product into the industry through a small firm, while early teething troubles are overcome. On the other hand, where a number of firms offer different brands of the new process, there may be greater pressure on them to effect an early prestigious sale to one of the giants in the adopting industry.

In the event, there was no evidence for the second hypothesis, but some confirmation for the first and third.

Define as S_j^* the size of the firm responsible for innovating process j (deflated by the mean size of firm in the adopting industry), NS_j as the number of suppliers of the new process, and C_j as the five-firm concentration ratio in the adopting industry. Then the following regression equation was computed:

$$S_j^* = 5.52 - 10.91\ C_j + 0.318\ NS_j, R^2 = 0.342.$$
$$(4.24)\quad\ (0.137)$$

(Estimated standard errors are shown in parentheses.) In other words, the innovating firms tend to be relatively smaller (1) the more concentrated is their own industry and (2) the fewer firms there are selling the process (a rough measure of the degree of competition between suppliers).

However, this (and the other results in this section), must be viewed in the light of a small and perhaps unrepresentative sample. The need for further research and data collection in this area is only too obvious.

Conclusions

This chapter has implicitly concerned the hypothesis that large size and small numbers of firms are conducive to rapid technical change. A number of conclusions have been drawn which are of some relevance to that hypothesis.

1. Typically, large firms adopt new processes more rapidly than small firms. There are strong technical reasons for this, and consequently it has not been suggested that this implies greater progressiveness by large firms.

2. An analysis of two theoretical models of diffusion suggests at least three potential influences of concentration on the speed at which an industry diffuses new processes. While two of these influences suggest an inverse relation between concentration and diffusion speed, the third influence suggests the opposite.

Diffusion, Innovation, and Market Structure

Empirically there is evidence that diffusion is slower in industries characterized by high size inequalities (and thus higher concentration), but in one study an inverse relation was also observed between diffusion speed and firm numbers. If these results are accepted, it is unlikely that a simple monotonic relationship exists between concentration and diffusion speed.

3. On the basis of a small sample of processes, there is evidence that large firms did not perform their share of innovating (especially of "major" processes). Moreover, innovators tended to be smaller in more concentrated industries.

I should add the qualification to the third conclusion, particularly, that further work is called for to see whether these findings can be confirmed on a wider basis.

Notes

1. See for instance, Kennedy and Thirlwall (1972), Kamien and Schwartz (1975), Weiss (1971), each of whom has, necessarily, little to report on concentration and diffusion.

2. Throughout this chapter the term *innovator* refers to the first firm to employ a new process within its own plants. In general, this will not be the producer or supplier of the process.

3. See, for instance, Romeo (1975), Globerman (1975), Nabseth (1973), Metcalfe (1970), Smith and Hakonson, both in Nabseth and Ray (1974). See also Davies (1979, chap. 2.) for some doubts on the statistical methodology employed by Mansfield and various of these authors.

4. The parameters of this function do, of course, change over time, such that the function shifts continuously in a north-westerly direction.

5. This convenient convention is seemingly well founded. For instance, Mansfield (1961) shows that once β is known, we may calculate the time lapse between diffusion attaining any two given levels. Similarly, $1/\sigma_D$ fulfills much the same role in the two alternative curves 8.3 and 8.4.

6. See Griliches (1957), Mansfield (1961), Romeo (1977, 1975), Globerman (1975), and many others.

7. The nature of Romeo's data perhaps leaves some room for doubt concerning the generality of this result, which is taken from a regression equation with only four degrees of freedom. Note also the narrow industrial base of his sample.

8. This algebraic formulation clearly acknowledges that an adopter at some time prior to t may not still employ the process at t. See Davies (1979, chap. 4) for a fuller discussion.

9. A more intuitive derivation of this result is possible, employing figure

8-3. Clearly, diffusion (the shaded area) will increase more rapidly over time (a) the more rapidly S_{ct} approaches the origin (the larger is g) and (b) the more "compressed" is the size distribution (the smaller is σ_s^2).

References

Adams, W. and Dirlam, J., 1966. Big steel, invention, and innovation. *Quarterly Journal of Economics* 80:161-189.

Davies, S. 1979. *The diffusion of process innovations.* Cambridge: University Press.

Globerman, S. 1975. Technological diffusion in the Canadian tool and die industry. *Review of Economics and Statistics.* 57:428-434.

Griliches, Z. 1957. Hybrid Corn: an exploration in the economics of technological change. *Econometrica.* 25:501-522.

Kamien, M., and Schwartz, N. 1975. Market structure and innovation: a survey. *Journal of Economic Literature.* 23:1-37.

Kennedy, C., and Thirlwall, A. 1972. Surveys in applied economics: technical progress. *Economic Journal.* 82:11-72.

Mansfield, E. 1961. Technical change and the rate of imitation. *Econometrica* October. 29:741-766.

──── . 1963a. The speed of response of firms to new techniques. *Quarterly Journal of Economics.* 77:290-311.

──── . 1963b. Size of firm, market structure, and innovation. *Journal of Political Economy.* 71:556-576.

──── . 1968. *Industrial research and technological innovation.* New York: Norton.

Mansfield, E., Rapoport, J., Schnee, J., Wagner, S., and Hamburger, M. 1971. *Research and innovation in the modern corporation.* New York: Norton.

Metcalfe, J. 1970. Diffusion of innovation in the Lancashire textile industry. *Manchester School.* 38:145-162.

Nabseth, L. 1973. The diffusion of innovations in Swedish industry. In *Science and technology in economic growth,* ed. B. Williams. New York: Wiley.

Nabseth, L., and Ray, G., eds. 1974. *The diffusion of new industrial processes.* Cambridge: University Press.

Romeo, A. 1975. Interindustry and interfirm differences in the rate of diffusion. *Review of Economics and Statistics.* 57: 311-319.

──── . 1977. The rate of imitation of a capital-embodied process innovation. *Economica.* 44:63-69.

Weiss, L. 1971. Quantitative studies of industrial organisation. In *Frontiers of Quantitative Economics,* ed. M. Intriligator. Amsterdam:North Holland.

Williamson, O. 1965. Innovation and market structure. *Journal of Political Economy.* 73: 67-73.

Technological Progress and Policy

Devendra Sahal

It is generally agreed that technological change has been the *élan locomotif* of progress in many spheres of human activity in the past. As an illustration, during the late fifties a number of studies of long-term economic growth reached a remarkable conclusion: that between 80 and 90 percent of the observed growth of output per capita in the American economy could not be accounted for by increase in capital per person and must therefore be due to technical progress. In a classical investigation in this area, Solow attributed only 12.5 percent of the growth of per capita output in the United States during 1909-1949 to the use of capital and the remaining 87.5 percent to technical change [30]. A similar conclusion was reached by Abramovitz by means of a different approach to analysis of observed economic growth in the United States during the period 1869-1953 [2]. Since then a host of studies have shed further light on the consequences of technological change. The general conclusion that emerges from these studies is indisputable: technological decisions constitute a powerful instrument of policy and conscious change.

Much less is known about the causes of technological change. For example, a number of analyses since the early studies of economic growth have revealed that a large portion of the observed increase in productivity is attributable to economies of scale and improvement in the quality of the inputs employed [10]. These changes of scale and quality in turn have been made possible by innovative activity. However, relatively little is known about the determinants of technological innovation. In recent years a number of investigations have significantly contributed to our knowledge in this area. However, a clear theoretical treatment of the process of technological innovation is generally lacking. In the succinct statement of Nelson and Winter [16, p. 38]: "The prevailing theory of innovation has neither the breadth nor the strength to provide much guidance regarding the variables that are plausible to change, or to predict with much confidence the effect of significant changes."

This is a sufficient justification for this study, in which an attempt has been made to come to grips with the questions how and why the process of technological development occurs. This chapter brings together some of the results of my earlier works in this area [19-25]. It presents two theoretical propositions of technological change and provides an operationalization of these propositions in the form of certain testable models of technology. Further, it will apply these and a number of other theoretical propositions to a variety of cases of technological change in farm machinery, transportation equipment, and digital computers. The chapter also examines the relative importance of various long-term determinants of technological change. Finally, the results of these case studies, which have a number of policy implications, are discussed in the conclusion.

Two Propositions in a Theory of Technological Change

This section outlines two central propositions concerning the process of technological change. The first proposition has its origin in the notion of the progress function that experience plays an important role in the improvement of productivity. The phenomenon relatively is well known. One rather specific aspect of this was first observed in the aircraft industry: that direct labor input per airframe declined substantially as cumulative output went up [33]. Further studies in this area have found that the concept of a progress function is applicable to a number of activities, including manufacturing, maintenance, and startup operations in various industries, including basic steel, heavy equipment, shipbuilding, electric appliances, and petroleum refining (see Sahal [24] and references therein).

While the applications of the progress function concept have generally been restricted to improvement in the performance of a system embodying a given technology, it is generalizable to the case of improvement in the technology itself. Virtually every case study of the history of technology reveals that progress in this area occurs largely in the form of seemingly minor but steady changes in some basic form of machine design. Even the major innovations must undergo extensive modifications before their potential can be fully exploited [21, 31]. A new technology does not emerge "like Minerva from Jove's forehead." Rather its emergence is very much a matter of countless modifications of some earlier, less specialized design. In this process accumulated experience of a practical nature plays a crucial role. Success in engineering comes primarily from "getting one's hands dirty" rather than from conceptualizing alone. In this respect technological progress is fundamentally different from advances in pure sciences. Even the most carefully conceived blueprints seldom prove to be practicable at once. Rather they must be put through the mill and tried out several times before they can be made operational. Typically, the new designs tend to be unreliable, inefficient, and cumbersome. Moreover, their execution into new products generally requires special dies, patterns, jigs, and templates. The installation of these new devices is seldom possible without giving rise to severe bottlenecks in the production process. However, as experience is gained, it becomes possible to identify and extirpate the bugs. The work force becomes better adapted to the task at hand, and the management finds new ways and means for improving plant layout and scheduling of material, labor, and equipment. With similar processes taking place in the supply area, materials and equipment of considerably improved quality become available. All these factors significantly contribute to the development of technology.

In summary, technological progress is never a one-shot affair. Rather, it is an evolutionary process involving extensive efforts of an experimental nature. These considerations suggest what may be called the "learning by doing" hypothesis of technological change.[1] In this view success in technical problem solving depends on the existing stock of practical experience. The solutions thus found, in turn, provide a new dimension to previous know-how. In this way technological change propagates itself in a cumulative manner.

The operationalization of this proposition requires a suitable measure of experience. We shall consider this problem in some detail, since it has been a subject of considerable controversy in recent years [3-6, 8, 12, 19, 24]. The process of acquisition or experience involves both time and activity. Corresponding to these two dimensions, experience can be measured in terms of either cumulated production of cumulated years of relevant experience, for example.[2] The case for the use of time as an appropriate measure of experience is that not all the know-how is acquired through the production activity. The crucial factor may well be the time factor involved in making the necessary alterations in the production setup, the hiring of qualified technicians, vocational training of the labor force, and so on. In particular, suppose that once the relevant know-how is acquired, it becomes readily accessible to all the firms in a given industry through movement of personnel, imitation, and the like. Under these circumstances, the use of cumulated output as a measure of industry's experience is inappropriate since it has a large element of redundancy built into it. However, progress in engineering comes largely from the actual exposure to the production process. Further, there is no reason to suppose that the inventive ability is unevenly distributed across the various firms in the industry. Insofar as occurrence of inventions is essentially a probabilistic process, production activity of every firm in the industry is of equal importance. In summary, considering the origin of the technical know-how, we can see that the relevant measure of experience is in terms of cumulated output rather than in terms of time. Once the know-how is generated, its successful adoption is again a function of experience acquired in the production activity. This is true even in the most unlikely case where there are no obstacles to industrywide transmission of new discoveries. For one thing, changes in the product characteristics often require far-reaching changes in the production process. The linkage between the product and the process technology becomes increasingly intimate as technology advances and acquires a system's characteristics. Thus successful adoption of a technology is generally conditional on changes in the production process. All in all, it seems that the variable, cumulated production quantities, is generally preferable to cumulated years of production as a measure of relevant experience in an analysis of the process of technological change.

While experience acquired in the production process plays an important role in the development of new technology, the fruits of experience can seldom be harvested as soon as they are acquired. For example, in the early development of the farm tractor, the Fordson model was produced in increasing quantities, as it continued to dominate the market for a period of nearly ten years, from 1918-1928. Yet during the same period its technical specifications remained virtually unchanged. The experience acquired in the production of Fordson proved to be extremely valuable in the long run rather than immediately [21]. Further many of the innovations that originate in the experience of the past often force a cutback in the current level of production because new manufacturing methods may be required that cannot be introduced at once. A good example is provided by the considerable delay in the delivery of the new 360 series of IBM computers a decade ago. There are, of course, many more examples of this phenomenon [1]. Their lesson is clear: the association between technology and experience essentially constitutes a long-term relationship.

The second proposition of this study has its origin in the observation that the development of a technology is generally governed by the conditions surrounding its use. A good example is the farm tractor. Consider the three main types of technology: the track type, the wheel type, and the garden variety. We may ask why it was necessary to develop more than one type of tractor in the first place. The answer appears to be that the development of the different types was necessitated by the differences of a topographical nature. Whereas the requirements of small farms and terrain of a more or less regular nature favored the wheel type of tractor, it was unsuitable for use elsewhere. Thus it was necessary to develop the crawler type of tractor for use on large farms and uneven terrain. Requirements of very small and much less specialized farms could be met only by the development of the garden type of tractor. The specific changes in the design of each type of tractor are again attributable to the conditions surrounding its use. The incorporation of the high rear-axle clearance and adjustable front spacing in the wheel tractor was an inevitable consequence of the practice of row crop cultivation. The progressive reduction in the size of the drive wheels, leading to the use of rubber tires in tractors, was likewise necessitated by the requirements of weak soils. Similarly, the incorporation of the four-wheel drive was necessitated by poor conditions for traction, as in the case of wet soils. Thus the design of today's tractor has been shaped to a large extent by considerations of a technical nature. This is equally true of the technology employed in other industries.[3] For example, many of the important changes in the transport technology are attributable to the characteristics of the available routes and stations. Thus improvement in characteristics such as speed and horsepower-to-weight ratio of road transport vehicles would have not been possible without concomitant improvement in the road standards and designs. In summary, the long-term development of technology is governed by the characteristics of the larger system of its use.

Conversely, the failure of technological development to take place under many circumstances is often due to lack of adaptation of the machine to the conditions of its use. Historically, the mismatch between technology and the larger system of its use has caused much irreparable havoc. A striking example of his is provided by the change of the irrigation system from the old basin method to perennial irrigation and the change of crop rotation system within perennial irrigation during the British occupation of Egypt in the years 1882-1914 [18]. The older system, dating back to the time of the Pharaohs, consisted of the canals dug through the highland along the banks of the Nile, while the remainder of the land was divided into basins by dikes. The canals enabled control of the flood water, while the dikes trapped the sediment-laden water. Under this system, not only was there little use for fertilizers, since the flood supplied basins with nutrients every year; very little land preparation was needed. Further, the occasional floding of higher fields as well as annual washing of the basin lands prevented the deposit of salt in the soils.

The new perennial irrigation made it possible to raise the level of water in the Nile and in canals during the dry preflood summer months by means of various dikes, dams, and canals. This enabled both expansion of the arable land and its intensified use by the switch from a three-year to a two-year crop rotation. However, the benefits that came from increased crop yields proved to be temporary at best, and the long-term consequences were nothing short of disastrous. The soil began to deteriorate rapidly, due to shortening the fallow, and the rise in the water table led to the suffocation of the deep roots of the plants and to soil salination. The intensified use of the land further contributed to rapid growth of insect pests leading to frequent and severe attacks.

Such examples of the adverse consequences of technological change are all too frequent. In modern times the "green revolution" has failed to materialize in certain parts of the world because the high-yield varieties have not been adapted to conditions where there has been lack of water control and little diversity of crops [14]. Likewise, transport technology has failed to take root in many developing countries because the basic infrastructure—the system of roads and waterways and the communication network—is not fully developed. All this is not entirely unexpected, since development of a technology is dependent on how well it dovetails with the system of its use.

In its influence on the process of technological change, the most salient characteristic of the system is its size. It is generally the case that the specialization of many of the vital activities of a system depends on its scale. The division of the labor in cities in terms of both the number and diversity of service establishments, manufacturers, and retail stores depends on their sizes measured in terms of their populations [34, p. 376]. Likewise, one finds that the number of occupational specialities and the number of occupational types in a so-called primitive society are governed by the size of the settlement [15]. A priori, it is not implausible that the process of technological development is a function of the size of the organization designed to secure its utilization. There are several reasons for this. First, the very adoption of a technology might depend on a certain threshold size below or above which its use may not be optimal [6]. Second, insofar as changes in size are generally accompanied by change in the number of operations, they may enable the use of an advanced technology. Indeed, certain changes in size may even require the development of a new technology. Finally, insofar as replacement requirements of a large-scale organization differ from those of small-scale counterparts, this may have a bearing on the process of technological change. It seems appropriate to regard these considerations in terms of what may be called a "specialization via scale" hypothesis of technological change. There is no presumption here that advances in technology require a certain "bigness" or "smallness" of the scale of organization for its use. On one hand, the use of a technology may be feasible only after a certain increase in the scale. On the other hand, the large scale of an organization may be a barrier to advances in technology insofar as its use may require extensive changes in the (existing) specialized equipment and plant. The important point is that changes in technology tend to be generally associated with change in the size of the organization designed to secure its utilization.

The operationalization of this proposition requires a suitable measure of the size of the organization in the use of technology. While the notion of the size of a system has been extensively employed in a variety of disciplines, including sociology, economics, and biology, there has been very little conceptualization of what it is. The term has been used in such widely different ways that it has come to resemble a black box in virtually every field (for a survey of the relevant literature, see Dullemeijer [7] and Kimberley [13]). My own attempt at measurement of size is as follows. First, it seems that the essence of the theoretical proposition can be best captured when size of the organization is defined in terms of the physical scale of operations of the technology under consideration. Second, an analysis of changes in different types of technology is best conceived in terms of different measures of the scale of operations. That is, different aspects of scale are primarily relevant for different types of technology. There is no universally applicable measure of scale. Our proposal then is as follows. The relevant measure of scale in considering the changes in agricultural technology pertains to some areal dimension of the physical organization such as average acreage per farm. In contrast, an appropriate measure of scale in considering the changes in transportation technology is a distance factor or some related variable such as route miles or number of ports. Yet other measures of scale must be employed in considering other types of technology. For example, one relevant measure of scale in considering the changes in the communications technology is the volume of transactions involved, for example, number of telephone calls made in a given period of time. In summary, different measures of scale would be appropriate to different situations.

I have argued that the changes in technology are determined by changes in the scale of some larger system of its use. However, is it not so that changes in size are in turn determined by changes in technology? The answer is that, to a certain extent, the relationship between the two variables may well be of a causal nature. However, in any analysis of technological change, it suffices to consider the size of the larger physical organization as a predetermined variable, for several reasons. First, the distribution of physical organization by size cannot change quickly. For example, technology is only one factor in determining farm size. Other factors include topography and types of crops produced. Frequently, the physical organization is a product of evolution over a much longer period of time in comparison with technology. Very often, it is neither possible nor desirable to alter the organization, especially when it is easier to change the technology. Thus today's tank ships follow many of the same routes used in the past. This allows them to utilize, when possible, many of the same natural phenomena that assisted navigation in ancient times: the southwest monsoon and its drift, the Mozambique and Agulhas currents, and the like. This in turn often enables considerable savings in fuel and time. Further, new canals and waterways cannot be built just because the technology may be available. The existence of technology may well be a spur to expansion of navigational facilities. However, the facilities cannot be expanded at will because there exist numerous constraints such as those posed by geography and demography. In summary, the determinants of the scale of organizations tend to be very numerous, and the effect of each rather small. For the purpose of analysis, therefore, it is justified to assume that causality runs from scale of the physical organization to the type of technology employed. Second, changes in different types of scale are primarily relevant for different aspects of technological change. The adaptation between technology and the scale of organization may indeed be of a mutual nature. However, this may well be confined to different dimensions. In certain dimensions at least the relationship between the two is of a one-way nature. Finally, time lags often involved in the way in which technology and the scale of organization of its use affect each other. When we consider the state of technology at a given point in time, the size of the organization at the same point is then an independent variable. All in all, the treatment of the scale of the organization as a predetermined variable in an analysis of technological change is not entirely without justification.

In summary, two theoretical propositions of long-term technological change have been advanced here. One has its focus on the history of technology. The other has its focus on the scale of operation of technology. The first proposition is essentially concerned with the role of time in the process of technological development, where time is measured in terms of a certain intrinsic property of the system. The second proposition seeks to bring about the role of space in this process. This point, while it may be obvious, deserves some attention. Hitherto, developmental studies have been concerned mainly with the role of time in the phenomenon under consideration. In their emphasis on dynamic aspects of the process, they have often ignored the equally important role of space. The two propositions advanced here attempt to correct this imbalance. Taken together, they imply simply that a developmental process takes place in time as well as over space. Further, while the first proposition attempts to elucidate the role of the machine-building sector, the second proposition seeks to clarify the role of the machine-using sector in the process of technological change. In summary, these two propositions of technological development are essentialy complementary.

Formal Specification of the Theoretical Propositions

The formal specification of the proposed theory rests on two fundamental features of the process of technological development [25]. First, the process of development must eventually reach a stage of equilibrium, because every system of a given form is subject to a limit of its growth. As an example, an apartment tower is necessarily limited in size because beyond a certain height the elevator space is bound to occupy a prohibitively large part of the total space. Similarly, growth of a power plant or a telephone exchange is ultimately limited by the increase in complexity with increase in size. Second, however, any given stage of a stagnation of a technology often proves to be of a temporary nature. One important reason for this is that in the evolution of technology, unlike the evolution in the organic world, two systems often combine in an integrative relationship. The formation of such a relationship often simplifies the overall form of the system such that each of the members in the coalition can continue to grow. A good example is the development of the three-point hitch and control system which made a thoroughly simplified system of tractor and implement, thereby rescuing both technologies from stagnation [21]. There are, of course, numerous other examples of this phenomenon.

The equilibrating aspect of technological change can be formalized in many different ways. One formulation is to consider the well-known Gompertz function

$$Y = K \exp\left[-\exp(A - Bt)\right] \quad (9.1)$$

where Y is a measure of technology, K is an equilibrium stage of growth, and A and B are constants. One central implication of 9.1 is that the relative growth rate $(1/Y) \times (dY/dt)$ decreases exponentially with time, since

$$dY/dt = KB \exp(A - Bt) \exp\left[-\exp(A - Bt)\right].$$

$$= BY \exp(A - Bt)$$

or $(1/Y)(dY/dt) = B \exp(A - Bt)$. \quad (9.2)

Further, from (9.1)

$$\log Y = \log K - \exp(A - Bt). \tag{9.3}$$

From (9.2) and (9.3)

$$d \log Y/dt = B (\log K - \log Y) \tag{9.4}$$

which can be estimated in the form

$$\log Y_t - \log Y_{t-1} = B (\log K_t - \log Y_t) \tag{9.5}$$

In view of the disequilibriating nature of development, the equilibrium level of growth, K in this formulation cannot be assumed to remain constant in the long run. According to the two theoretical propositions advanced earlier, the long-term development of technology (consisting of variations in K) is determined by the accumulated production experience (X), and the scale of a relevant larger organization (Z) of its use, respectively. One simple way to express these relationships is in the forms

$$\log K_t = \alpha_1 + \beta_1 \log X_t \tag{9.6}$$

and

$$\log K_t = \alpha_2 + \beta_2 \log Z_t. \tag{9.7}$$

Two principal formations of the process under consideration are obtained by combining equation 9.5 with 9.6 and 9.7, respectively:

$$\log Y_t = \alpha_1 (1-\lambda) + \beta_1 (1-\lambda) \log X_t + \lambda \log Y_{t-1} \tag{9.8}$$

and

$$\log Y_t = \alpha_2 (1-\lambda) + \beta_2 (1-\lambda) \log Z_t + \lambda \log Y_{t-1}, \tag{9.9}$$

where

$$\lambda = 1/(1+B) \quad \text{and} \quad 0 < \lambda < 1, \tag{9.10}$$

since B is essentially a positive quantity. According to these formulations, the coefficient itself of any given explanatory variable, say X, is an estimate of its effect on the chosen measure of technology Y in the short run. The estimate of the long-term effect of X on Y can be obtained by dividing the observed coefficient of X by $1-\lambda$ where λ is the coefficient of the lagged dependent variable. Finally, this form of specification is applicable to any hypothesis postulating an essentially long-term relationship between technological change and its determinants.

Case Studies of Technological Change

This section presents a number of case studies of technological change in aircraft, tank ships, locomotives, farm tractors, and computers over the course of time. Specifically, the following cases of technological change are examined.

1. a. Average fuel consumption efficiency of farm tractor in horsepower-hours per gallon, 1920-1968.
   b. Average mechanical efficiency (the ratio of drawbar horsepower to belt horsepower) of farm tractor, 1920-1968.
   c. Average belt horsepower per 1,000 lbs of unballasted tractor weight, 1920-1968.
2. Average tractive effort of locomotives in lbs, 1904-1967.
3. Average service speed of tank ships in knots, 1914-1970.
4. Average speed of commercial aircraft in miles per hours, 1932-1965.
5. a. Average performance of computers for scientific use in operations/sec, 1943-1967.
   b. Average performance of computers for commercial use in operations/sec, 1943-1967.

The chosen measures of technology and the explanatory variables employed in the case studies have been dictated by empirical necessity. A detailed discussion of the qualitative aspects of the various cases of technological change under consideration is available in my earlier works and will not be repeated here [20-23]. However, a few observations on their scope are in order. To begin with, the chosen measures of technology enable one to take into account both major and minor innovations and assign appropriate weights to their importance according to a certain common denominator. An example will help make this clear. A number of innovations ranging from frameless or unitary form of construction, hardened steel gears, and removable cylinder liners to development of power takeoff and the use of pneumatic tires have contributed to the evolution of farm tractors. It is not easy to assign dates to most of them. For example, it took nearly two decades to exploit the full potential of the power takeoff from the time of its first introduction. The relatively perfected form of the three-point hitch and control system developed in 1938 was an outcome of more than seventeen years of experimental effort, starting with the development of an integral tractor plow. These specific examples point to what is generally true: virtually every innovation has to undergo countless modifications over the course of time before the desired degree of performance and reliability can be

Technological Progress and Policy

secured. In consequence, any attempt to assign a specific date to the origin of an innovation is bound to be unreliable. Further, numerous difficulties arise in distinguishing between an innovation and a minor variation in technology. And even if a list of innovations can be agreed on, assigning weight to each of them according to some chosen measure of their importance remains a Herculean task. In short, any conceptualization of a technology in terms of the number of relevant innovations is fraught with numerous difficulties. However, there is an alternative: when advances in the technology are conceived in terms of a functional property of the machine such as horsepower-to-weight ratio or fuel consumption efficiency, not only does it become possible to take into account the various underlying innovations, but they are also automatically weighted according to their contribution to an objectively measurable characteristic of the phenomenon under consideration. In this manner this study bypasses many of the sins of omission and commision which have hitherto made objective studies of innovation all but impossible.

Test of the "Learning by Doing" Hypothesis

This section is concerned with an application of the learning-by-doing hypothesis to various cases of technological innovation under consideration. With the possible exception of the tank ship, a priori there is considerable evidence to indicate that technological change in these cases is attributable largely to the initiative of equipment manufacturers rather than the equipment users. For example, consider the case of technological change in aircraft. There are perhaps no more than eight major airlines that are powerful enough to dictate the design and performance specifications of a new type of aircraft. In general, the orders placed by individual airlines do not affect the basic design [32]. Consequently, the relevant variable in the explanation of technological change is the experience accumulated in the machine-producing rather than the machine-using sector. This is particularly true of the other cases of technological change under consideration, namely those concerning the agriculture and the service industries in which the suppliers of the equipment are relatively few while the users are numerous and incapable of very cohesive action. In summary, one would expect technological change in virtually all the observed cases to have its origin in the experience acquired by the machine builders.

This expectation is borne out by the results presented in table 9-1. In seven of nine cases the theoretical model leaves a mere 1 to 8 percent of variance unexplained in the data. The relationship between technological change and cumulated production is shown here in a diagrammatic form for the illustrative case of the farm tractor (figure 9-1). It is equally good in other cases although the qualitative aspects, of course, differ from case to case. Except where there

Table 9-1
Parametric Estimates of the Learning-by-Doing Formulation in Various Cases of Technological Innovation

Case	Technology	Time Period	Dependent Variable (y)	Independent Variable (x)	Estimated Relationship	R^2	S	F	N	$D.W.$	Source
1a	Farm tractor	1921-41 1948-68	Average fuel consumption efficiency in hp-hr/gal	Cumulated tractor production	$\log y_t = -0.05 + 0.07 \log x_t + 0.68 \log y_{t-1}$ (0.08) (0.04) (0.12)	0.93	0.03	247.6	42	2.24	23, 25
1b	Farm tractor	1921-41 1948-68	Average mechanical efficiency (ratio of drawbar hp to belt hp)	Cumulated tractor production	$\log y_t = 0.42 + 0.05 \log x_t + 0.64 \log y_{t-1}$ (0.14) (0.02) (0.12)	0.89	0.02	153.84	42	1.74	23
1c	Farm tractor	1921-41 1948-68	Average belt horsepower per 1000 lbs of unballasted tractor weight	Cumulated tractor production	$\log y_t = -0.07 + 0.09 \log x_t + 0.56 \log y_{t-1}$ (0.09) (0.03) (0.14)	0.76	0.05	61.52	42	1.85	19, 23
2a	Locomotive	1904-45	Average tractive effort in pounds	Cumulated locomotive production	$\log y_t = 0.17 + 0.0016 \log x_t + 0.96 \log y_{t-1}$ (0.03) (0.005) (0.01)	0.99	0.004	19,333.8	42	1.65	22
2b	Locomotive	1904-57	Average tractive effort in pounds	Cumulated production of freight and passenger cars	$\log y_t = 0.46 + 0.056 \log x_t + 0.91 \log y_{t-1}$ (0.04)	0.99	0.02	6,696	54	1.84	22
3	Tank ship	1914-70	Average service speed in knots	Cumulated tank ship production	$\log y_t = 0.007 + 0.003 \log x_t + 0.99 \log y_{t-1}$ (0.01) (0.002) (0.01)	0.99	0.005	7,093.3	57	1.35	22
4	Aircraft	1947-65	Average air speed in mph	Cumulated additions of new aircraft	$\log y_t = -0.11 + 0.007 \log x_t + 0.99 \log y_{t-1}$ (0.09) (0.01) (0.05)	0.99	0.008	989.04	19	1.23	19, 22
5a	Computers for scientific use	1944-67	Average computing capability in ops/sec	Cumulated computer production	$\log y_t = -1.75 + 2.43 \log x_t + 0.17 \log y_{t-1}$ (0.73) (0.71) (0.22)	0.95	0.75	86.56	23	1.86	20
5b	Computers for commercial use	1944-67	Average computing capability in ops/sec	Cumulated computer production	$\log y_t = -1.61 + 2.48 \log x_t + 0.07 \log y_{t-1}$ (0.69) (0.68) (0.22)	0.92	0.82	60.30	23	1.97	20

Note: R^2 is the coefficient of determination, S the standard error of estimation, F the ratio of variance explained by the model to the unexplained variance, $D.W.$ the Durbin-Watson statistic, and N the total number of observations. Standard errors of estimated coefficients are given in parentheses.

Technological Progress and Policy 183

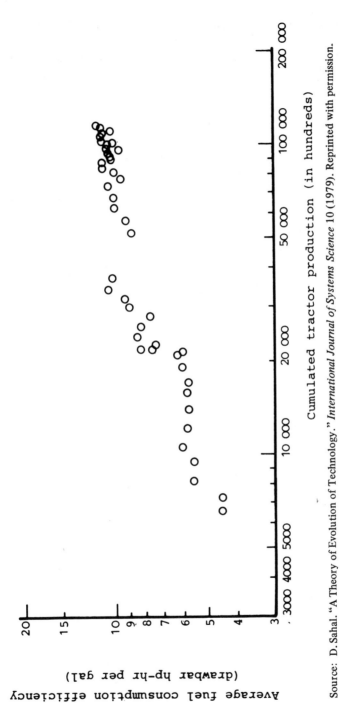

Source: D. Sahal. "A Theory of Evolution of Technology." *International Journal of Systems Science* 10 (1979). Reprinted with permission.

Figure 9-1. Relationship between Fuel Consumption Efficiency of Tractors and Cumulated Tractor Production, 1920-1968

are problems due to multicollinearity, the parametric estimates of the variables are highly significant.[4] The coefficients of the lagged dependent variable are quite high in all three cases of change in transportation technology. In part, this is due to the high correlation between the explanatory variables themselves. For the remainder, it reflects the fact that the trend of growth in these cases is very nearly exponential. Because of the presence of lagged dependent variables in the estimated equations, the Durbin-Watson test statistics are biased and should not be regarded as evidence against correlation in the residuals. Its estimate here is presented as a test for presence rather than absence of serial correlation. (This interpretation of the Durbin-Watson test applies throughout to all the results of this study and not just those presented in table 9-1). In this regard the performance of the model meets the required criterion at $\alpha = 0.01$ level in all cases except that of aircraft and tank-ship technology, where the results are inconclusive. But even in these cases, R^2 values are so high that the confidence in the theoretical model is justified.

In summary, the results do not refute the theoretical proposition. They further indicate that the role of learning is not confined to improved performance in the use of a technology. The development of technology itself is also a function of learning. That is, the role of learning in the improvement of productivity is potentially unlimited.

Test of "Specialization via Scale" Hypothesis

The parametric estimates of the model based on the specialization-via-scale hypothesis are presented in table 9-2 for various cases of technological innovation on which the data are available. Equations 1a-1c correspond to the various aspect of technological change in farm tractors. The coefficient of the farm size variable in these equations is generally insignificant because it is highly correlated with the lagged dependent variable. The relationship between technological change and farm size is shown here in a diagrammatic form for the illustrative case of the tractor (figure 9-2). It holds equally well in other cases although the qualitative aspects differ from case to case. At the $\alpha = 0.01$ level of significance the results of the Durbin-Watson tests are highly satisfactory in all the cases considered. The performance of the model is generally good. The conclusion that emerges from these results is that over the course of time the manufacturers introduced various changes in tractor technology so as to meet the requirements imposed by change in the farm size. Equation 2 provides a test of the theoretical proposition in case of changes in the locomotive technology during the period 1904-1967. The coefficient of the total railroad mileage is insignificant because it is highly correlated with the lagged dependent variable. The Durbin-Watson test value meets the desired criterion quite well, and the

Table 9-2
Parametric Estimates of the Specialization-via-Scale Formulation in Various Cases of Technological Innovation

Case	Technology	Time Period	Dependent Variable (Y)	Independent Variable (X)	Estimated Relationship	R^2	S	F	N	$D.W.$	Source
1a	Farm tractor	1921-41, 1948-68	Average fuel consumption efficiency in hp-hr/gal	Average acreage per farm	$\log y_t = 0.015 + 0.05 \log x_t + 0.86 \log y_{t-1}$ (0.09) (0.06) (0.07)	0.92	0.03	228.7	42	2.49	23, 25
1b	Farm tractor	1921-41, 1948-68	Average mechanical efficiency (ratio of drawbar hp to belt hp)	Average acreage per farm	$\log y_t = 0.21 + 0.048 \log x_t + 0.82 \log y_{t-1}$ (0.10) (0.042) (0.09)	0.87	0.02	136.29	42	1.82	23
1c	Farm tractor	1921-41, 1948-68	Average belt horsepower per 1000 lbs of unballasted tractor weight	Average acreage per farm	$\log y_t = -0.36 + 0.37 \log x_t + 0.37 \log y_{t-1}$ (0.14) (0.10) (0.15)	0.79	0.05	73.72	42	1.83	23
2	Locomotive	1904-67	Average tractive effort	Total railroad mileage	$\log y_t = 0.11 + 0.009 \log x_t + 0.97 \log y_{t-1}$ (0.23) (0.045) (0.01)	0.99	0.01	4,484.1	64	2.31	22
3a	Aircraft	1933-65	Average air speed	Number of airports and landing fields	$\log y_t = 0.03 + 0.02 \log x_t + 0.96 \log y_{t-1}$ (0.06) (0.02) (0.05)	0.98	0.01	825.9	33	1.18	22
3b	Aircraft	1933-65	Average air speed	Total route miles	$\log y_t = 0.01 + 0.003 \log x_t + 0.99 \log y_{t-1}$ (0.06) (0.03) (0.06)	0.99	0.01	804.5	33	1.18	22

Note: R^2 is the coefficient of determination, S the standard error of estimation, F the ratio of variance explained by the model to the unexplained variance, $D.W.$ the Durbin-Watson statistic, and N the total number of observations. Standards of estimated coefficients are given in parentheses.

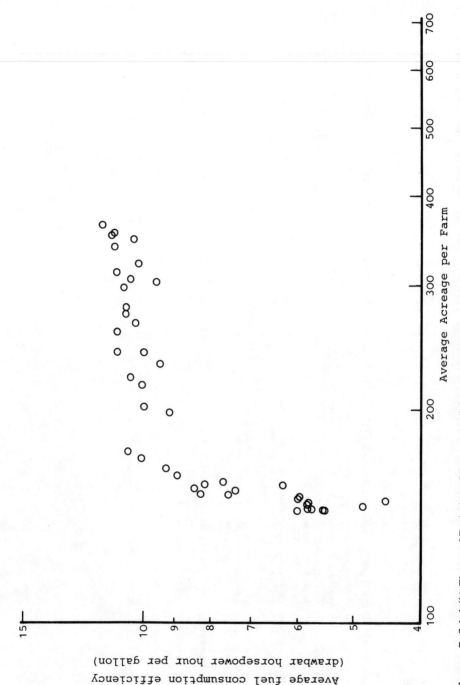

Source: D. Sahal. "A Theory of Evolution of Technology." *International Journal of Systems Science* 10 (1979). Reprinted with permission.

Figure 9-2. Relationship between Fuel Consumption Efficiency of Farm Tractors and Average Farm Size 1920-1968.

model leaves a mere 1 percent variance unexplained in the data. Equations 3a and 3b correspond to an explanation of change in average aircraft speed during the years 1933-1965 where the scale of operations is measured in terms of number of airports and landing fields and total route miles. The latter variable, of course, is a measure of many variables besides the scale; it reflects the joint influence of a number of factors including the stock of aircraft, demand for air travel, and maintenance. In both cases the coefficients of the explanatory variables are insignificant due to problems of multicollinearity. Further, the results of the Durbin-Watson test are inconclusive. However, as can be seen from the R^2 values, the predictive performances of both equations are still very high. In summary, the results do not refute the a priori evidence that the choice to buy a new aircraft (or locomotive) is inevitably based on the characteristics of the airline's (or the railroad's) routes. At an aggregated level, then, the evolution of transport technology is guided by the characteristics of the available routes and the inauguration of new routes.

Test of "Learning via Diffusion" Hypothesis

It is evident that the two main hypotheses of technological change considered thus far represent somewhat extreme views regarding the origin of the product innovation. According to the learning-by-doing hypothesis, innovations originate in the machine-building sector. The specialization-via-scale hypothesis, on the other hand, postulates the origin of innovations in the machine-using sector. The former assumption is generally supported by the a priori information regarding the cases of product innovations examined in this chapter. However, it is plausible that the problems are first pinpointed in the machine-using sector and that their solutions are then incorporated in the subsequent versions of technology. That is, while innovation is indeed very much a matter of initiative taken by the machine builders, the feedback obtained from the users of the new technology plays an important role in this process. According to this viewpoint, the relevant variable in an explanation of technological change is cumulated utilization of technology (the stock of machines in use) rather than cumulated production quantities. These considerations may be regarded in terms of what may be called a learning-via-diffusion hypothesis of technological change. In its essence, this hypothesis regards the origin of technological change in between the two extremes previously considered. The meaning of it is very simply that the diffusion of technology paves the way for improvements in its characteristics.

The formal specification of this hypothesis is somewhat problematic in that the system may be potentially simultaneous. Thus changes in the characteristics of a technology often lead to new uses of the technology, thereby significantly affecting the course of its diffusion. A single-equation model is nevertheless justified insofar as there exist considerable time lags in the system of these relationships. The form of the model chosen here is identical to that of the models previously considered.

The parametric estimates of the model in various cases of product innovation are presented in table 9-3. As indicated by the values of the coefficients of determination and the results of the Durbin-Watson test, the model performs quite well in its explanation of the various aspects of technological change in farm tractors over the course of time. Its performance is less satisfactory, however, in its explanation of change in various types of transport technology. In the case of the locomotive the improvement in the tractive effort turns out to be negatively related to cumulated utilization of technology measured in terms of the existing stock of locomotives. While such a result is not implausible in that the existing stock may be an obstacle to adoption of innovation, it does not support the basic hypothesis under consideration. For the remaining equations, the standard errors of the estimated coefficients are generally very low. The significance of the estimated coefficients is admittedly exaggerated in the case of both tank-ship and aircraft technology where the results from the Durbin-Watson test are inconclusive. Nevertheless, it is fair to say that the parametric estimates of the equations are generally well determined and that the explanatory power of the model is very high in virtually all cases of technological change under consideration.

Alternative Hypotheses

In an attempt to further determine the origin of learning in the process of technological change, we have also considered an alternative hypothesis of the disadvantage of being the *fons et origo*. In its essence, pioneers in the field of technological development suffer a disadvantage relative to the newcomers for a variety of reasons, ranging from resistance to change to the effect of sunk costs [9]. Thus it has been frequently asserted that industrialized countries, such as Britain, are handicapped in their attempt to modernize the industry due to their early start. Clearly, the hypothesis is diametrically opposed to the learning-via-diffusion hypothesis of technological change.

In its operational form, the hypothesis of the disadvantage of being the *fons et origo* may be taken to mean that the process of technological change is retarded by the age of the existing capital stock. That is, the lower the age of the capital stock, the better the prospects for technological progress. The age variable can be readily measured as a ratio of capital stock to gross investment.

The results from the test of the hypothesis are presented elsewhere [26]. The hypothesis was found to perform poorly in virtually every case of technological change with the possible exception of locomotive and tank ship. Nonetheless, the sign of the age variable was positive (and therefore theoretically incorrect) only in the case of changes in the fuel consumption efficiency and

Technological Progress and Policy

Table 9-3
Parametric Estimates of the Learning-via-Diffusion Formulation in Various Cases of Technological Innovation

Case	Technology	Time Period	Dependent Variable (y)	Independent Variable (x)	Estimated Relationship	R^2	S	F	N	D.W.	Source
1a	Farm tractor	1921-41, 1948-68	Average fuel consumption efficiency in hp-hr/gal	Number of tractors on farm	$\log y_t = 0.04 + 0.07 \log x_t + 0.69 \log y_{t-1}$ (0.04) (0.04) (0.12)	0.93	0.03	246.1	42	2.25	23
1b	Farm tractor	1921-41, 1948-68	Average mechanical efficiency (ratio of drawbar hp. to belt hp)	Number of tractors on farm	$\log y_t = 0.46 + 0.05 \log x_t + 0.66 \log y_{t-1}$ (0.16) (0.02) (0.12)	0.89	0.02	151.6	42	1.75	23
1c	Farm tractor	1921-41, 1948-68	Average belt horsepower per 1000 lbs of unballasted tractor weight	Number of tractors on farm	$\log y_t = 0.05 + 0.09 \log x_t + 0.56 \log y_{t-1}$ (0.07) (0.04) (0.14)	0.76	0.05	61.83	42	1.85	23
2	Locomotive	1904-67	Average tractive effort in pounds	Number of locomotives in service	$\log y_t = 0.17 - 0.002 \log x_t + 0.96 \log y_{t-1}$ (0.14) (0.017) (0.01)	0.99	0.01	4,481.8	64	2.32	22
3	Tank ship	1900-73	Average service speed in knots	Stock of tankship	$\log y_t = 0.007 + 0.004 \log x_t + 0.98 \log y_{t-1}$ (0.008) (0.002) (0.01)	0.99	0.004	13,848.7	74	1.25	22
4	Aircraft	1933-65	Average airspeed in miles per hour	Number of aircraft in service	$\log y_t = 0.07 + 0.029 \log x_t + 0.93 \log y_{t-1}$ (0.06) (0.013) (0.03)	0.99	0.01	935.05	33	1.16	22

Note: R^2 is the coefficient of determination, S the standard error of estimation, F the ratio of variance explained by the model to the unexplained variance, D.W. the Durbin-Watson statistic, and N the total number of observations. Standards of estimated coefficients are given in parentheses.

mechanical efficiency of the farm tractor. In the remaining four cases of changes in the horsepower-to-weight ratio of the tractor, tractive effort of the locomotive, service speed of the tank ship, and speed of the aircraft, the age variable was found to enter with a negative sign in accordance with the theoretical proposition under consideration. It may therefore be concluded that technological change in agriculture appears to have been due to learning in both machine-building and machine-using sectors, while innovations in transportation appear to have been made possible almost wholly through learning in the machine-building sector.

We have also tested yet other propositions in cases where the data are available: the demand-pull, technology-push hypothesis that innovative activity in any given industry varies in direct relation with the volume of its sales and the theory of induced innovations that the firms are forced to search for new labor-saving techniques as labor becomes dearer relative to capital.[5] However, according to the results presented elsewhere [26], the performances of both hypotheses were found to be distinctly poor in comparison with either the learning-by-doing or specialization-via-scale hypotheses of technological change.

Relative Importance of Long-Term Determinants of Technological Change

The formal specification of the theoretical proposition presented in this chapter readily enables one to estimate both the short-term and the long-term influence of chosen variables on the process of technological change. Thus the long-term effect of a variable, say x on the chosen measure of technology, say y, can be obtained by dividing the observed coefficient of x by $1-\lambda$ where λ is the coefficient of lagged dependent variable. The resulting estimates of the long-term effect of learning and scale on technological change are presented in tables 9-4 and 9-5.

The results indicate that the estimated coefficients of learning in technological innovation generally differ from case to case. In turn, the coefficients of learning are generally lower than those of scale in comparable cases of technological change.

It is of considerable interest to determine whether the observed differences in the estimated long-term coefficients are significantly different.[6] This can easily be done by means of analysis of variance. The results are presented in tables 9-6 and 9-7. At the $\alpha = 0.05$ level, the difference between the learning coefficients associated with the development of farm machinery and transportation technology is not significant. This is somewhat expected since developments in both areas have been at least partly made possible by advances in mechanical engineering. Further, we have included aircraft and tank ships as well as locomotives in our sample of learning coefficients associated with

Technological Progress and Policy

Table 9-4
Long-Term Role of Learning in Various Cases of Technological Innovation

Case	Source (table 9-1)	Technology	Measure of Technology	Long-term Coefficient of Learning
1	1a	Farm tractor	Fuel consumption efficiency	0.22
2	1b	Farm tractor	Mechanical efficiency	0.14
3	1c	Farm tractor	Horsepower to weight ratio	0.20
4	2a	Locomotive	Tractive effort	0.04
5	3	Tank ship	Service speed	0.30
6	4	Aircraft	Air speed	0.70
7	5a	Computers for scientific use	Computing speed	2.93
8	5b	Computers for commercial use	Computing speed	2.67

Source: Table 9-1.

advances in transport technology. While such a classification of farm machinery and transport technology is justified on an industrywide basis, it is quite arbitrary from a strictly technological standpoint. According to the latter, farm tractor and locomotive constitute one group, while tank ship and aircraft constitute another group. This is because advances in the former group of technologies have their origin in the experience acquired in mechanics, while advances in the latter group have been due to experience gained in the subject of transport (hydrodynamics or aerodynamics). The results presented in table 9-6 indicate that there exist significant differences between the two groups of learning coefficients when they are classified on a technological basis as defined before. The results further indicate that long-term coefficients of learning in the

Table 9-5
Long-Term Role of Scale of Operations in Various Cases of Technological Innovation

Number	Case (table 9-3)	Technology	Measure of Technology	Long-Term Coefficient of Scale
1	1a	Farm tractor	Fuel consumption efficiency	0.36
2	1b	Farm tractor	Mechanical efficiency	0.27
3	1c	Farm tractor	Horsepower to weight ratio	0.59
4	2	Locomotive	Tractive effort	0.30
5	3a	Aircraft	Air speed	0.50

Source: Table 9-2.

Table 9-6
Significance of Differences in the Long-Term Role of Learning in Various Cases of Technological Innovation

Industry or Technology		Mean Value of Long-Term Coefficient of Learning	F-Matrix ($v_1 = 1, v_2 = 5$)		
Industry-based classification					
Agriculture	(y_1)	0.187		y_1	y_2
Transportation	(y_2)	0.347	y_2	0.74	–
Computer	(y_3)	2.8	y_3	158.62	139.79
Technology-based classification					
Tractor and locomotive	(y_1)	0.15		y_1	y_2
Tank ship and aircraft	(y_2)	0.50	y_2	6.12	–
Computing devices	(y_3)	2.80	y_3	350.35	198.27

Source: Table 9-4.

case of computer technology are significantly different from those in case of either farm machinery or transportation technology regardless of how they are classified. In summary, we have the important result that the role of learning tends to be technology specific.

Table 9-7 presents a comparison of the estimated coefficients of learning and scale in technological change in agriculture and transportation industry, respectively. In either case, at the $\alpha = 0.05$ level, the estimated F-ratio is insignificant, thereby indicating that the long-term coefficients of scale do not significantly differ from those of learning. Thus it appears that the role of learning in comparison with scale differs from case to case.

Two important conclusions emerge from the analysis presented in this section. First, there exists a significant difference in the susceptibility of different types of technology to improvement from learning. Second, in many

Table 9-7
Relative Importance of Long-Term Role of Learning and Scale in the Process of Technological Change

Learning and Scale Coefficient	Mean Value
Farm sector	
Long-term coefficient of learning	0.187
Long-term coefficient of scale	0.407
F-ratio ($v_1 = 1, v_2 = 4$)	5.750
Transportation sector	
Long-term coefficient of learning	0.346
Long-term coefficient of scale	0.400
F-ratio ($v_1 = 1, v_2 = 3$)	0.040

Source: Tables 9-4, 9-5.

cases, learning plays as important a role as scale in the process of technological innovation.

Conclusions and Policy Implications

This chapter has presented a theory of technological change. Two principal hypotheses have been outlined. One is the learning-by-doing hypothesis, that technological change grows out of experience. The other is the specialization-via-scale hypothesis that technology is a function of the scale of the organization of its use. It is found that a wide variety of cases of technological change in farm machinery, transportation equipment, and digital computers can be adequately explained by means of either of the two hypotheses. Clearly, the implication is that technological progress is intimately linked with the process whereby the work force acquires the production skills and whereby the relevant organization for the use of technology is developed. Seen in this light, an explanation of the hitherto elusive residual factor in long-term economic growth turns out to be relatively simple.

On the basis of the results from studies of such widely different cases of technological change, as in farm tractors, aircraft, tank ships, locomotives, and computers, it may be said that the role of learning is particularly important to the growth of productivity. The relevance of learning extends far beyond improvement in the use of a technology. The origin of technology itself is also a function of learning. Thus one explanation of long-term economic growth is to be found in terms of what might be called an economywide learning, whereby the production skills are acquired over the course of time. There are a number of salient features of this process. First, a large proportion of growth in productivity due to changes in technology employed in many sectors of economy is demonstrably attributable to the experience acquired in the production activities in some other sectors. According to the results of this study, while technological change in farm machinery is attributable to learning in both the machine-building and the machine-using sectors, technological change in transportation appears to have taken place almost solely through learning in the machine-building sector. That is, the origin of a variety of technological innovations can be traced to the experience acquired in the machine-building rather than in the machine-using sector. We have not conclusively shown that the converse is false. What we have shown is that the process of learning does take place in the machine-building sector. Insofar as data are available, there is a strong presumption in favor of the hypothesis that (cumulated) gross investment in any given sector of the economy is an essential vehicle whereby its productivity is improved. This confirms what has long been suspected but never adequately proved: to a certain extent, all investment has the character of investment in R&D.

Second, since the estimated long-term coefficient of learning differs from case to case, it may be concluded that the role of learning tends to be technology specific. That is, different types of technology exhibit very different scopes of advances resulting from learning. This has the important theoretical implication that technology shares the characteristic of putty-clay capital (putty ex ante, clay ex post). In practical terms this means that initial choice of a technology is of supreme importance. Investing in a technology is like giving hostage to fortune, for it will continue to exert influence in the course of subsequent development. The consequences of an initial choice, that is to say, cannot be easily repealed.

Third, one further related consequence of the differences in the observed learning coefficients is that results of learning in the development of a technology are not entirely transferable to development of some other technology. If they were, there would be a tendency toward equalization of the long-term coefficients of learning. However, as we have seen, this is far from being true. Thus a new type of technology cannot be developed overnight from the skills acquired in the development of other technologies in the past. This has the important policy implication for the future that training of the labor force toward acquisition of alternative types of skills will be increasingly necessary with the growing need for new technologies due to scarcity of resources.

The results of this study also shed some light on the causes and cure of underdevelopment. The process of technology transfer from industrialized countries to developing economies has seldom succeeded in achieving its goals. According to the results of this study, the process is indeed doomed to failure from the start. First, technological progress generally takes place in close connection with the process whereby skills are acquired and the necessary organization for the use of technology is developed. Mere installation of a technology is therefore necessarily inadequate. Rather it has to grow. Seen in this light, the prevailing policy of transfer of ready-made technology is something of a contradiction in terms. An appropriate form of technology transfer is instead transfer of the relevant recipes rather than the bakeries themselves. Second, if the results of this study are any guide, the roles of both learning and scale of organization tend to be highly specific to the type of technology involved instead of facilitating technological development across the board. Consequently, the advanced state of skill and organization in the industrialized countries may have little bearing on the type of technology suited to the needs of the developing countries. The implication is that the developing countries will have to undertake independent R&D efforts of their own. Further, as shown in the study, all investment is ipso facto a kind of research, and consequently investment in R&D alone is inadequate. In conclusion, it is suggested that an appropriate policy in this area should be aimed at some form of transfer of investment rather than transfer of technology per se.

The results of this study also have a bearing on the important issue of tariffs and other forms of subsidies aimed at infant industry prorotection. It has been suggested that some form of protection is justified in order to increase the opportunities through which learning and technological progress take place over time. The results of this study suggest that any such policy prescription is subject to a number of qualifications. First, a wide variety of innovations are attributable to learning in the machine-building sector. Consequently, in many instances, any protection offered to the machine-using sector is tantamount to penalizing the machine-building sector of the economy. The appropriate policy for both developed and developing countries in such cases is, instead, one of eliminating the obstacles involved in the way of industrywide diffusion of know-how originating in the machine-building sector. Second, learning is not the only factor involved. According to the results of this study, development of organization required for the use of technology is no less a crucial consideration. For example, some form of subsidy may be justified until the scale of organization is fully developed in the use of technology. Last but not least, in view of the fact that the roles of both learning and scale of organization tend to differ widely in the case of different types of technology, what might be justifiably considered for subsidies is not the industry. It is, instead, a specific technology.

Notes

1. The expression "learning by doing" was coined by Arrow [3]. The term *learning* in this expression is a misnomer, since it is not the same as the process under consideration—acquisition of relevant experience. The term is nevertheless employed here because of its widespread use in the literature on the subject.

2. The controversy surrounding the use of these alternative measures is somewhat misguided. These measures differ largely in their substantive rather than in their formal characteristics. At least in certain cases, there is no formal, theoretical difference between these two types of measures. To begin with, consider the commonly specified form of *unit* progress function

$$y_i = a_1 (x_i)^{b_1} \qquad (1)$$

where, for example, x_i is the cumulative total of units produced and y_i is a measure of performance of the system. As shown elsewhere [24], equation 1 is equivalent of the cumulative density function of the Pareto distribution

$$F(u) = P(U < u) = 1 - \beta u^{-\alpha} \qquad (2)$$

where $P(U < u)$ is the probability that a random variable U assumes a value less than u and α and β are constants, such that $b_1 = -1/\alpha$ and $a_1 = (N\beta)^{-b_1}$ where

N is the total number of observations. Next consider, for example, the following form of the *average* progress function

$$\bar{y}_i = a_2 (t_i)^{b_2} \qquad (3)$$

where $\bar{y}_i$ is the average level of performance during the ith interval of time and t_i is a measure of time such that the successive intervals are equidistant. Once again the finding holds. On the other hand, the following type of average progress function

$$\bar{y}_i = a_3 (\bar{x}_i)^{b_3} \qquad (4)$$

cannot be interpreted in the same way as 1 and 3. Nevertheless, insofar as all three forms hold, they are representative of certain invariant aspects of evolutionary systems.

3. For a similar viewpoint see Solo [29] and Reynolds [17].

4. There are printing errors involved in the parametric estimates of the relationships concerning changes in the horsepower-to-weight ratio of the tractor and in speed of aircraft as reported in Sahal [19]. The corrected forms of the relationships are shown here. Further, the relationships concerning changes in computing capability differ slightly from those reported in Sahal [20]. This is because the latter employ natural logarithms, while the former are based on the use of common logarithms.

5. An authoritative exposition of the theory of demand of technology can be found in the pioneering work of Schmookler [28]. It may be criticized on the ground that while demand may indeed be a stimulus for innovation, the solutions of technical problems cannot be found at will.

The theory of induced innovations was originally set out by Hicks [11]. In its essence, technical advances inherently tend to have a labor-saving bias. One main objection against this viewpoint, as noted by Salter, is that aim of the firm using a technique is to minimize the total costs, and it is immaterial whether this is accomplished by saving labor or capital [27].

6. The question arises to what extent the observed differences are due merely to the effect of differences in the units of the various measures of technology and the explanatory variables. According to the modern theory of dimensions, the relevant issue is not one of true dimensions, but rather which choice of dimensions will be of maximum utility. It seems appropriate to regard all the measures of technology as well as the cumulated production quantity variable in terms of the dimension of time [24]. Further, the scale variables employed in the models of technological change in farm machinery and transportation equipment are best regarded in terms of the dimensions of area and length, respectively. Consequently, while significance of observed differences in the long-term coefficients of scale can be determined only in

comparable cases (where the same measure of scale is employed), there does not seem to be any restriction in the assessment of long-term coefficients of learning.

References

[1] Abernathy, W.J., and Wayne, K. Limits to the learning curve. *Harvard Business Review*, Sept.-Oct. 1974, pp. 109-119.

[2] Abramovitz, M. Resource and output trends in the U.S. since 1870. *American Economic Review, Papers and Proceedings* 46 (1956):5-23.

[3] Arrow, K.J. The economic implications of learning by doing. *Review of Economic Studies* 29 (1962):155-173.

[4] Arrow, K.J. Classificatory notes on the production and transmission of technological knowledge. *American Economic Review, Papers and Proceedings* 59 (1969):29-35.

[5] Atkinson, A.B., and Stiglitz, J.E. A new view of technological change. *Economic Journal* 79 (1969):573-578.

[6] David, P. *Technical choice, innovation, and economic growth.* Cambridge: University Press, 1975.

[7] Dullemeijer, P. *Concepts and approaches in animal morphology.* Assen: van Gorcum, 1974.

[8] Fellner, W. Specific interpretations of learning by doing. *Journal of Economic Theory* 1 (1969):119-140.

[9] Frankel, M. Obsolescence and technological change in a maturing economy. *American Economic Review* 45 (1955):296-319.

[10] Griliches, Z. The sources of measured productivity growth: U.S. agriculture 1940-60. *Journal of Political Economy* 71 (1963):331-346.

[11] Hicks, J. *Theory of wages.* London: Macmillan, 1932.

[12] Ishikawa, T. Conceptualization of learning by doing: a note on Paul David's "Learning by doing and ... the ante bellum United States cotton textile industry". *Journal of Economic History* 33 (1973):851-861.

[13] Kimberley, J.R. Organizational size and the structuralist perspective: a review, critique, and proposal. *Administrative Science Quarterly* 21 (1976):571-597.

[14] Myint, H. *South-east Asia's economy.* Harmondsworth: Penguin, 1972.

[15] Naroll, R.S. "An index of social development. *American Anthropologist* 58 (1956):687-715.

[16] Nelson, R.R., and Winter, S.G. In search of useful theory of innovation. *Research Policy* 6 (1977):36-76.

[17] Reynolds, L.G. Discussion. *American Economic Review, Papers and Proceedings* 56 (1966):112-114.

[18] Richards, A. Technical and social change in Egyptian agriculture, 1890-1914. *Economic Development and Cultural Change* 26 (July 1978).

[19] Sahal, D. A reformulation of technological progress function. *Journal of Technological Forecasting and Social Change* 8 (1975):75-90.
[20] Sahal, D. The relevance of the logic theory machine to modelling of evolutionary systems. *Kybernetes* 6 (1977):49-53.
[21] Sahal, D. Evolution of technology: a case study of farm tractor. Berlin: International Institute of Management, 1978.
[22] Sahal, D. Models of technological development and their relevance to advances in means of transportation. Berlin: IIM, 1978.
[23] Sahal, D. Farm tractors and the nature of technological innovation. Berlin: IIM, 1978.
[24] Sahal, D. A Theory of progress functions. *Transactions of the American Institute of Industrial Engineers* 11 (1979).
[25] Sahal, D. A theory of evolution of technology. *International Journal of Systems Science,* 10 (1979):259-274.
[26] Sahal, D. Recent advances in a theory of technological change. Berlin: International Institute of Management, 1979.
[27] Salter, W.E.G. *Productivity and technical change.* Cambridge: University Press, 1969.
[28] Schmookler, J. *Invention and economic growth.* Cambridge, Mass.: Harvard University Press, 1966.
[29] Solo, R. The capacity to assimilate an advanced technology. *American Economic Review, Papers and Proceedings* 56 (1966):91-97.
[30] Solow, R. Technical change and the aggregate production function. *Review of Economics and Statistics* 39 (1957):312-320.
[31] Usher, A.P. *A history of mechanical inventions.* Cambridge, Mass.: Harvard University Press, 1954.
[32] Williams, J.E.D. *The operation of airliners.* London: Hutchinson, 1964.
[33] Wright, T.P. Factors affecting the cost of airplanes. *Journal of the Aeronautical Sciences* 3 (1936):122-128.
[34] Zipf, G.K. *Human behavior and the principle of least effort.* Cambridge, Mass.: Addison-Wesley, 1949.

Part III
Science and Technology Policy

The papers in part II examined a number of implications of the process of innovation for technological planning. The papers in part III are concerned largely with implications of technological planning for the process of innovation. In particular, the objective of their inquiry is to understand the mechanisms involved in shaping the course of technological change to meet certain desired goals. Studies in part I attempted to address this issue at the level of a firm. Studies in this part attempt to do the same at the level of an industry or in a national context.

The chapter by Bruun is concerned with the process of technology transfer, the adaptation of technology to alternative tasks. The process of technology transfer is, of course, closely related to the process of diffusion of technology, examined in part II; the distinction between the two is relative rather than absolute. Diffusion involves adoption of a new technology for a given use, while transfer involves adoption of a given technology for new uses. Thus the process of technology transfer generally involves changes in the technology itself. Depending on the nature of these changes, two types of technology transfer are generally distinguished in the literature on the subject: the horizontal transfer of technology to fit alternative job descriptions and the vertical transfer of technology from the more general to the more specific. The chapter by Bruun is concerned with the ways and means to enhance the pace of horizontal transfer of technology developed for space and defense purposes. The substantive basis of his study is a survey of nearly one hundred new production centers in the Aarhus county in Denmark, which together employ about two thousand people. One of the most important factors in the growth of these firms is the know-how acquired by their founders in their previous occupation, usually at some firm in private industry rather than in a government research institution. The founders of a minority of new technology-based firms (those previously employed by government agencies) are typically well educated but lacking in entrepreneurial skills. Similar findings have been reported by studies in other countries. The implication is that courses in business education can go a long way toward promoting technological spin-off from the public sector. Further, the largest group of new firms has its origin in the direct transfer of technology from the parent institutions. Significantly, while this group of firms got started with an established product, as well as a market, they contributed most to the improvement of old products and to the development of new ones. One finds, therefore, that the successful instances of technology transfer are characterized by a certain continuity which could be promoted by government research agencies and those concerned with policy planning.

The chapters by Schwarz and by Pelc present a detailed examination of the nature of technological planning. Schwarz's work focuses on the process of planning; Pelc's work emphasizes the product of planning. The two chapters offer an interesting contrast between the institutional settings in that one concerns technologically advanced countries and the other concerns developing countries. The starting point of both papers is that the problems of technological planning are best pursued in the larger context of the industrial sectors involved. One important conclusion of Schwarz's work is that there exist considerable differences between the various sectors of the economy, both with regard to the demands on the design of the planning process and the problems of implementation. Further, the relationship between planning and research activity works both ways. It is not just a case of usefulness of planning for orienting the research activity: the basic issue is as much one of usefulness of research for planning. Technologists and policy planners ought not to be going their own ways. Rather a close cooperation between the two is essential to the success of a long-term technology strategy.

In contrast with this process approach, Pelc outlines a structural approach to technological planning. It is an attempt to analyze the problem from more than a competitive market viewpoint. Two alternative criteria of technological planning emerge from the analysis. One criterion has to do with the intersectoral differences in the state of technology. Moreover, there exist significant interactions between various sectors, particularly between the leading and the lagging areas. This makes it necessary for any changes in policy to be introduced simultaneously in the various sectors of the economy. In effect, such a strategy is based on a consideration of the range of implementation of technological advances. Explicit measurement of technological change provides a second criterion for planning. In effect, an appropriate strategy is to take calculated risks in bypassing the intermediate stages of technological development, in favor of explorations of the frontiers of knowledge.

The relevant knowledge has to do with technology per se rather than formally expressed knowledge as, for example, in the natural sciences. This point is forcefully brought home in the chapters by Fores and by Price. In the past technological policy has been all too frequently regarded as synonymous with applied science policy. According to Fores and Price, however, nothing could be more misleading. Technology is seldom an outcome of science; rather science is expanded as an outcome of technology. The notion of applied science does not take one very far, and it may even be misleading. A more important concept is that of bottom-up technology transfer in contrast with top-down or vertical transfer of technology. Bottom-up transfer may be regarded as the transfer of relevant know-how from the more specific to the more general, from technology to science. For example, there is considerable evidence to indicate that the contribution of the steam engine to thermodynamics far exceeds the contribution of thermodynamics to the steam engine. There are, of course, numerous other examples of this phenomenon, the reasons for which are not difficult to see. Advances in technology are made possible by empirical rather than formally expressed, exact types of knowledge. The principal dynamic force behind technical progress is the acquisition of relevant production skills rather than advances in theoretical knowledge. Moreover, a great majority of innovations depend on the development of an existing technique. The situation is neatly summed up by Price as the case of an "almost technology: something which we know roughly speaking, how to do (in principle) but which has not yet been put into practice." Much technical progress is essentially a transition from an almost to a complete technology. It is evidently made possible by the production system more often than by pure scientific knowledge. To put it succinctly, science and technology are two very different animals. The point is not merely semantic. Rather, it has profound implications for policy. Investment in the acquisition of production skills is no less crucial to technical progress than is investment in formal R&D activity.

10 Technology Transfer and Entrepreneurship

Michael O. Bruun

In the last twenty years the industrialized countries have invested considerable amounts in large research and development projects. The size of the amounts can be measured in percentage of the gross national product (GNP) [1, 2]. The investments have taken place mainly within the technical and natural sciences resulting in the buildup of advanced technology mainly of an electronic and electromechanical nature. [3, 4].

Since the beginning of the sixties there has been an increasing interest in the application of the know-how or knowledge behind this technology. The question asked is, To what extent has a technology transfer taken place, influencing the supply of goods and services on the civil (commercial) market (improvements of existing products or processes or new products or processes)?

The main arguments in favor of this interest are made on the basis of the well-established (but not fully understood) connection between technical change and economic growth [5, 6 and references herein]. Rasmussen [6] includes a very interesting postulate, that the present economic recession could be partly described (explained?) as part of a "Kondratieff cycle" generated in the postwar period by the chemical and electronic sectors of industry. The "seas of basic knowledge" within these fields are drying out. An increase in the rate of technology transfer of the kind mentioned before will counteract this tendency.

The key questions are now, Is there a substantial amount of useful technology, which at the moment is not commercially applied, and if the answer to this question is yes, How do we improve on the chances, that this relevant technology is transferred to the commercial market?

A technology transfer of the kind discussed can take place either through the foundation of a new company or through the transfer of know-how or knowledge to an existing firm. In the first case technicians acting as entrepreneurs are typically involved, whereas in the latter case the company may have obtained the knowledge in connection with an industrial contract including the development of the technology in question or from the literature or through engaging one or more of the involved technicians.

Transfer to an Existing Firm

A necessary condition for technology transfer is that the company involved must be able to absorb the technology in question, that is, there must be a match between the technical capabilities of the contractor and the contracting institution. (Transfer to an existing firm is illustrated in figure 10-1.) If this condition is met, the first step in the process is made possible (the technology buildup). The transfer of new technology or know-how in the firm to a process or product on the commercial market represents the most critical step of the process. To quote Edwin Mansfield [8], "The commercial risks—the risks that the new product will not be profitable in the market—tend to be much larger than the technical risks in civilian, R and D." The willingness or motivation to take this risk, including the necessary marketing efforts (and often further development), is strongly dependent on the anticipated profitability and on the market (the need). Add to this the size of the competitive pressure weighing on the company [9]. In accordance with this, in an investigation of the recent innovations in housing, railroads, and computers Myers and Marquis found that technological innovations are more often stimulated by perceived production and marketing problems and needs than by technological opportunities [10]. It is claimed (see for instance Pessemier [11]) that within a broad variety of industrial sectors between 60 percent and 80 percent of all important innovations are based on ideas generated in response to a market need. These results are supported by Donald A. Schon in his enlightening discussion of the possible interaction between the marketing function and the R&D function of the firm [12].

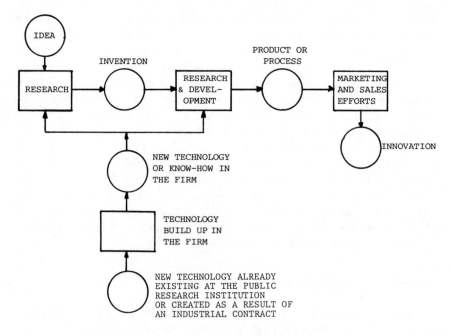

Figure 10-1. The Technology Transfer Process

What all this means is that even if a technology buildup has taken place, it is more than doubtful whether it ever will pass through the activities shown in figure 10-1 and end up as part of an innovation (or improved product or process) on the marketplace. In addition, the companies successful in obtaining the contracts involving new advanced technology are also characterized by a marketing function with a weak or nonexistent involvement on the commercial market.

The technology can be transferred in one of the following three ways: (1) through literature (paper, reports), (2) through the exchange of co-workers, or (3) through industrial contracts. In the early sixties American authorities already felt that the generation of technology was taking place at a much more rapid pace than did the later utilization for products or services "useful for society." In particular, the investments in aerospace technology were already at an early stage accompanied by promises of tangible payoffs. As a result of a corresponding political pressure, NASA established the Technology Utilization Program. The program was directed toward an increase in the technology buildup through new literature (mechanism 1).

Besides the more traditional ways of information dissipation (distribution of regular journals, reports), the Technology Utilization Program consisted of the Regional Dissimination Center Program. The latter involved the establishment of a number of active libraries around the country. The program has so far not been very successful, and few innovations survived the procedure shown in figure 10-1 [7, 13].

Nothing has so far been done to encourage the exchange of co-workers between the public research institutions and the relevant companies, and not much is known about the extent to which such exchanges actually take place.

In the case of an industrial contract there is little doubt that the technology buildup is more substantial than in the two other cases. Investigations seem to indicate that the main part of the advanced industrial contracts involves either the very large multiproduct, multitechnology companies or the very small scientifically based companies [14, 15, 7]. In both cases the chance that the technology might "get stuck" in the company is quite large.

Although investigations by Roberts and his co-workers [16] seem to show that the small scientifically based companies still have an entrepreneurial approach and thus should be specially fit to take part in a technology transfer process, they often lack the necessary marketing capabilities [7, 14, 15]. Marketing activities on the space-defense or research market are in many respects very different from "normal" marketing activities, and a small firm tends to concentrate its resources on only one of these lines of action. The small scientifically based companies were baptized "group-α" firms [7, 14, 15] and "Boston Route 128 spinoff firms" [16].

The large companies should in this respect be better equipped than any other company, but in these companies the necessary communication between the departments doing governmental work and the departments doing civilian work may be minimal. Further, even if the departments are doing both kinds of work, the specific item transfer seems to be very modest [17].

If the technology is stuck in a company, not much can be done except to increase the incentives to use that technology through a spin-off.

Entrepreneurship and Technical Change

I believe that one of the most promising ways to enchance the rate of technology transfer of the type discussed in this chapter is to stimulate or encourage the foundation of new independent spin-off companies. The corresponding business ideas include production on the basis of the know-how obtained by the entrepreneurs during their occupation either at one of the public research institutions or at one of the contract companies. It is obvious that the new companies will be of the group-α type, based as they are on the use of advanced technology. To get a deeper understanding of the factors influencing the possible start of a new company, I undertook an empirical investigation of the Danish entrepreneur [19].

This approach is of course more indirect compared to the one used by Edward B. Roberts and his colleagues in their investigation of the spin-off companies of several MIT laboratories and academic departments, a government laboratory, a nonprofit corporation, and an industrial electronic systems contractor [16]. Nevertheless I felt it important to be able to place the high-technology entrepreneur in the general picture of entrepreneurship.

The entrepreneur survey was carried out during 1976-1978 in the county of Aarhus in Denmark. One hundred and six production companies were selected and their entrepreneurs were interviewed. The criteria used in the selection were that the companies were operating in markets outside the local region and that they were founded either in the seventies or (as control groups) in 1950 or 1960. At the time of investigation the ninety-nine companies still operating employed about two thousand people.

From the survey I found that it was possible to describe every start as a specific interplay of four factors (or ingredients): (1) the educational and social background of the entrepreneur, (2) the occupational situation of the entrepreneur, (3) the business idea, (4) the resources (availability of capital). Figures 10-2 and 10-3 show the distribution of the sample on education (technical) received by the entrepreneurs and on the employment characteristics of the entrepreneurs' fathers. Three of every four entrepreneurs were either unskilled or skilled workers, and over half of the entrepreneurs (53 percent) had fathers who were self-employed; only 8 percent of the fathers were publicly employed.

Figure 10-4 shows the age distribution of the entrepreneurs at the time of starting the new enterprise. On average the new company founder is in his early thirties, but the distribution includes founders much younger and much older than that.

Comparing these characteristics of the ordinary (Danish) entrepreneur with the ones of the technical entrepreneur found in the study by Roberts [16], one

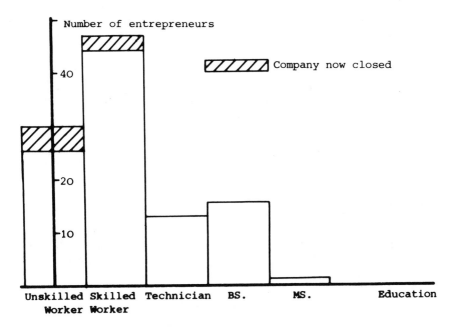

Figure 10-2. Distribution of Entrepreneurs by Education

finds that they are nearly identical except for the educational level. Not surprisingly, the level of education (technical) is considerably higher in "Roberts's" spin-off sample, with an average education slightly better than a master of science degree.

The distribution on figure 10-3 is explained in the same way as in Roberts. "During childhood the eventual entrepreneur has often had home influences of career orientation toward self-employment affecting his personal development." For the 47 percent of entrepreneurs without a self-employed father, in most cases it was possible to identify another self-employed person close enough to the entrepreneur to influence his career orientation and personal development.

Why people choose to set up their own company is of course one of the key questions in connection with entrepreneurship (entrepreneurial motivation). The detailed answer to this question involves factors 1 to 3, but the investigation also showed that it was possible to identify three main reasons for starting. Thus it is possible to divide the entrepreneurs into three main groups: (1) those who always wanted to become independent (34 percent); (2) those who were brought by chance or by accident into a situation where one of the alternatives was to start on their own (40 percent); (3) those who felt it was the only possibility in a situation, who for some reason could not continue in their previous job (26 percent). A contingency table using this classification of the entrepreneurs as one of the variables and father's occupation as the other supports the hypothesis that the two variables are independent (chi-square test).

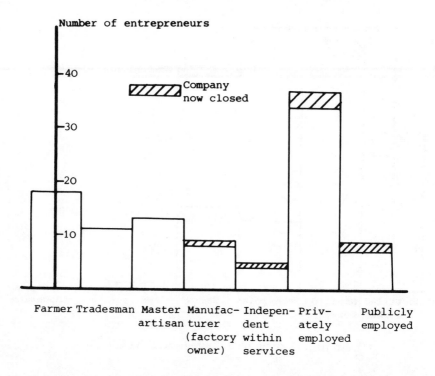

Figure 10-3. Distribution of Entrepreneurs by Father's Occupation

This finding can tentatively be interpreted as follows. A necessary condition for entrepreneurship is an entrepreneurial heritage from fathers (or relatives, collaborators, family friends) who were in business for themselves. This is a demand independent of the actual situation leading to the establishment of the new enterprise. The ability to see oneself as an independent business man is of vital importance.

In the two-way table shown as table 10-1 the reason for starting is again used as one of the variables. The other variable classifies the product produced as the start product (the technical part of the business idea). Only entrepreneurs from going concerns are included. Not surprisingly, most of the entrepreneurs (71 percent) used either an established product (taken from the mother company) or functions as their technical basis. Functions mean that the company is offering technical skills to the customers (for example electroplating or complex turning operations).

The chi-square test does support the hypothesis that the two variables are interdependent ($\alpha = 0.04$). This can be given the following interpretation.

1. The entrepreneur who always intended to set up his own business waits only until he feels he has a solid business idea and a chance to raise the necessary funds. It is of course most likely that he is exposed to such a complete business

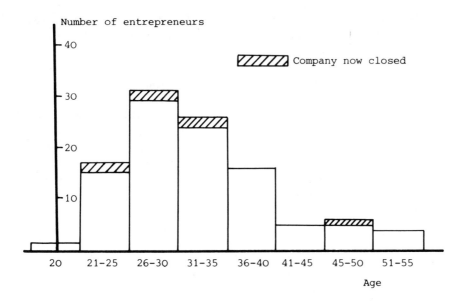

Figure 10-4. Distribution of Entrepreneurs by Age at Time of Startup

idea in the mother company, from which he then takes both technology and market idea. Corresponding to this only three (9 percent) of the entrepreneurs in this study started with an innovation.

2. The largest group consists of the entrepreneurs who are brought by outside forces or influences into a situation where one of the alternatives is to set up an independent company. Although the business idea directly transferred from the mother company is again dominating (58 percent), the tendency toward innovation is nearly twice the average. The group thus contains only 40 percent of the entrepreneurs but 67 percent of the innovations.

Table 10-1
A Classification of New Firms

	Reasons for Starting			
Initial Production	Always Intended to	One among Several Opportunities	Only Opportunity	Total
Function	10	3	1	14
Existing product	16	20	20	56
Changed product or process	4	8	3	15
Innovation	3	8	1	12
Miscellaneous	1	1	0	2
Total	34	40	25	99

The ones chosing the "well known" are typically displaced persons (the mother company has to close down), whereas the innovative entrepreneurs start either because of a perceived need or because of some disappointments or annoyances in connection with the development of the innovation in the mother company. The entrepreneurs were directly involved in this development and were thus able to transfer the innovation to the new company.

3. In the third group are the entrepreneurs without a choice. Because of age or skills not in great demand the entrepreneur felt that self-employment was the only alternative to unemployment. Nearly all these entrepreneurs chose the business idea from the mother company (84 percent).

To get a viable basis for their new enterprise, most of the entrepreneurs started by producing well-known products (or processes); they operated with an established technology as well as an established market.

For most of the innovative entrepreneurs the innovation played a major role in the course of events leading to the entrepreneurial decision. The innovation represented either an opportunity or a source of disappointment. In both cases the innovation was closely connected to a customer need perceived either by the entrepreneur himself or by the mother company.

Finally a company's starting on the basis of an existing product does not imply that it does not contribute to the technical development. Most of these companies had developed new or improved products at the time of interviewing.

The importance of the market side of the business idea is also stressed in connection with the performance of the technical entrepreneur. Roberts identified as a differentiating element "the existence of a marketing department in the more successful new companies and a lack of formal marketing organizations in the less successful firms," and he concludes "that this factor gives testimony that advanced technology needs actively to be exploited, and that even brilliant ideas do not move by their own energies into the market place" [16, p. 235].

The Founding of New Independent Companies

The distribution of the 106 companies on the different branches of industry is shown in table 10-2. Only the electronic companies and a few of the mechanical firms were based on high technology, which in that respect makes them comparable to the group-α companies [7] and the Route 128 companies [16].

Our empirical investigation now shows that the entrepreneurs behind these group-α like companies all belong to group 2 of the preceding section—none of them started because of a long-felt wish! In addition, these entrepreneurs can be characterized as loyal and competent employees who never would have thought of starting their own businesses if the circumstances had not suddenly made it a realistic alternative. The story in which the coming entrepreneur loses his job for the second time within a few years because his company or department again

Table 10-2
Distribution of the Sample on the Different Branches of Industry

Branch of Industry	Percentage of Sample
Textiles and clothing	11
Wood and furniture	16
Graphic	10
Metal and mechanical engineering	44
Electronic	6
Building and construction	7
Miscellaneous	6
Total	100

had to close down is a typical story. After this the entrepreneur felt the necessity to "control his own destiny." All the companies seemed to be well managed with a reasonable income and liquidity.

The group-α characteristics are the following:

The entrepreneur had a technical education well above average.

The business idea involved the use of high technology.

The company sells the main part of its production to public institutions.

The products on which the starting up of the company is based are more or less directly transferred from the mother company.

The more successful technical entrepreneurs Roberts brought a high degree of technology from the source laboratory to the new enterprise. The same holds for all the high-technology entrepreneurs of our sample; they all used the technology of the mother company. In both cases this choice is closely connected to the customer part of the business idea. The Route 128 companies had their starts as government contractors or suppliers to the defense and space market, whereas our high-technology entrepreneurs were selling their products on the commercial market. The difference arises when the Route 128 company eventually started selling on the commercial market. This might give rise to the innovation on the commercial market discussed in a preceding section.

Encouraging Entrepreneurial Activity in High Technology

We now turn to the question of increasing the inclination among the technical employees at the public research institutions or at the contract companies to start their own companies. The interest of these people to become independent manufacturers or businessmen is at first quite modest. In addition, they lack interest in and knowledge of marketing problems as well as the marketplace itself. This finding is in accordance with the findings of an American study of new technologically based firms by A.C. Cooper [18]. The aim of the study was to single out the type of organization that was most frequently the mother organization in the spin-off process. The study was carried out around 1970 in the Bay Area south of San Francisco. It is well known that the occurence of newly founded high-technology companies is quite large in this area. One of the main results of this investigation was that the numerous research institutions in the area (universities, governmental laboratories, nonprofit organizations) were seldom mother organizations. The situation for the large companies of the area was more or less the same, whereas the small companies had spin-off rates (number of new companies per employee) a number of magnitudes above.

One way to provoke more entrepreneurial activity in this atmosphere of low interest could be to make all the technical jobs at the public research institutions temporary, but this is probably not a very sound suggestion because of its negative influence on the performance of the technicians involved. Influencing the potential entrepreneurs through "education" seems to be a more promising solution. The results of one investigation [19] suggest that the wish to become independent or the tendency to start a company in an urgent situation is closely connected to one's ability to see oneself as an entrepreneur. The corresponding experiences have typically been obtained through the father's occupation as an independent businessman, craftsman, or farmer. The people from among whom we intend to recruit new entrepreneurs typically lack the natural entrepreneurial background and have educations at higher levels than was the case with the entrepreneurs of our survey.

The following proposal may thus be seen as an attempt to compensate for this lack of positive experience and for the lack of marketing experience. In order to stimulate the starting of new companies through the force of example I thus propose the introduction of courses on starting a company at the institutes of higher education. The courses might be framed according to the principles used at American universities, where field activities constitute the most important part of the course. Similar courses should also be offered occasionally during the later career of the potential entrepreneurs.

This could be combined with different means to make better conditions for small companies. One of the findings from the investigations of the impacts of the American space program is that "the willingness of space-defense procurement officials to purchase for instance semiconductor devices from small new firms often stimulated engineers and managers from established companies to set up their own firms" [3]. An increase in the exchange of publicly and privately employed technicians is desirable, because experience in trade and manufacturing is vital to an entrepreneur.

References

[1] OECD. *Survey of resources devoted to research and development by OECD member countries.* Paris, 1974.

[2] OECD. *Patterns of resources devoted to research and experimental development in the OECD area, 1963-71.* Paris, 1974.

[3] Schnee, Jerome E. Space program impacts revisited. *California Management Review* 20 (1977):62.

[4] Robbins, Martin D., Kelley, John A., and Elliot, Linda. Mission-oriented R. and D. and the advancement of technology: the impact of NASA contributions. Industrial Economics Division, Denver, Colo.: Denver Research Institute, 1972.

[5] Schmookler, Jacob. Invention and economic growth. Cambridge, Mass.: Harvard University Press, 1966.

[6] Rasmussen, P. Nørregaard. On growth and technological change. University of Copenhagen, 1976.

[7] Bruun, M.O. A Study of economic follow-effects caused by the industrial contracts of a large basic research center (with special emphasis on the technology transfer aspect). Institute of Management, University of Aarhus, 1977.

[8] Mansfield, Edwin. Technology and Technological Changes, in *Economic Analyses and the multinational enterprise*, ed. John H. Dunning. London: Allen and Unwin, 1975.

[9] Gruber, William H. The development and utilization of technology in industry. In *Factors in the Transfer Technology* ed. William H. Gruber and Donald G. Marquis. Cambridge, Mass.: MIT Press, 1969, p. 39.

[10] Myers, Summer and Marquis, Donald. Successful industrial innovations, Report to the National Science Foundation (NSF 69-17). Washington D.C., 1969.

[11] Pessemier, Edgar A. Product management strategy and organization. Pullman, Wash.: Washington State University Press, 1977.

[12] Schon, Donald A. *Technology and change.* London: Pergamon, 1967.

[13] Doctors, Samuel I. *The NASA technology transfer program.* New York: Praeger Publishers, 1971.

[14] Bruun, M.O. Economic "follow-effects" of manufacturing contracts placed in industry by a large basic research center. In *Proceedings of the Symposium on the Management of Research and Development.* Kiel University Press, 1975.

[15] Bruun, M.O. Some aspects of the transfer of technology from scientific applications to commercial applications. Presented at Seminar on Management of Research and Development, Brussels, April 1978.

[16] Roberts, Edward B. Entrepreneurship and technology. In *Factors in the transfer of technology.* ed. W.H. Gruber and D. Marquis. Cambridge, Mass.: MIT Press, 1969, p. 219.

[17] Black, Guy. The effect of government funding on commercial R and D. In *Factors in the transfer of technology.* ed. W.H. Gruber and D. Marquis. Cambridge, Mass.: MIT Press, 1969, p. 202.

[18] Cooper, Arnold C. Spin-offs and technical entrepreneurship. *IEEE Transactions on Engineering Management* EM-18 (1971):2-6.

[19] Bruun, Michael, Sørensen, Per, and Ravn, Niels. Danish entrepreneurs. Jutland Technological Institute, Denmark, 1978 (English summary).

11 Linking Research Planning to Sectoral Planning

B. Schwarz

In many European countries a characteristic of the development of national science policy in recent decades has been the increasing emphasis on the link with governmental policy in various sectors much as health, transportation, agriculture, defense, housing. In Sweden a generally held opinion in recent years has been that futures studies and long-range planning in the sectors should be closely linked to sectoral science policies. This was also the basic idea behind a recent transport study (Schwarz and Svensson 1978), which was initiated in 1975 by the Swedish Transport Research Delegation (TRD). This is a research council that commissions R&D projects but also funds R&D initiated by researchers. In the original mission description TRD called the study a "prospective plan." Its task was to provide a basis for R&D planning in the field of transportation.

The aim of this chapter is to throw some light on the relationship between long-range sectoral planning and R&D planning. In Sweden ideas about this relationship are directly related to the concept of sectoral research. To provide some background I shall therefore briefly describe the conceptual development from applied research to mission-oriented and sectoral research. I then describe the transport study. The emphasis here is on the conceptual model, a type of strategic planning model, that structured the study. Finally, experience from the transport study is used to discuss the usefulness of planning for orienting research as well as the usefulness of research for planning.

From Applied Research to Sectoral Research

There have been many changes in the last few decades in the organization and funding of research, particularly research of a more applied or problem-centered kind. Parallel to such factual changes have been developments in many of the concepts and ideas concerning science policy.[1]

In Sweden, as in many other European countries, the organization of research has generally followed a characteristic pattern: Fundamental research has been tied to the universities, while separate research institutes have been set up as the need for applied research in new areas has been recognized. In Sweden the establishment of separate research institutes started during the last century.

Some early cases were the Geological Survey of Sweden (1858) and the Swedish Meteorological and Hydrological Institute (1873). Examples of organizational developments during the first half of this century are the establishment of the National Veterinary Institute (1911), the Metallographic Institute (1920), the National Institute for Plant Protection (1932), the National Road Research Institute (1934), and the National Defense Research Institute (1945).

The 1950s and 1960s represented a period of considerable growth for both the universities and the separate research organizations. Committees were appointed from time to time to investigate the need for reforms or for new research institutes. However, in connection with discussions about such changes, the division of research into fundamental and applied research was found inadequate. Also the term *applied research* was felt to be somewhat misleading. It gives the impression that the separate research institutes are simply applying results obtained in fundamental research at the universities to specific practical problems. This is often not the case. Applied research is not necessarily based directly on previous research. Characteristically, it has its raison d'être in its potential contribution to real-world problems in given areas. To signify this aspect of applied research the term *mission*-oriented research (*målforskning, målinriktad forskning*) came to be used increasingly in Sweden during this period.

Mission-oriented research can be said to be product oriented, that is, aimed at developing material to meet a given specification; or it can be problem oriented, that is, aimed at finding an acceptable solution to some practical problem. When the potential contribution to real-world problems is not regarded as direct and likely to occur in the near future, then mission-oriented research can be very similar in character to fundamental research. In this case it may be termed mission-oriented basic research (*riktad grundforskning*).

The difference between mission-oriented research and traditional university research, which is usually termed fundamental or basic research but also known as pure or free research, does not necessarily concern the character of the work but is more a matter of its context. In fundamental research the subject of the research is chosen independently by the research workers themselves with no eye to utility or material productivity. In this choice criteria intrinsic to the discipline concerned are paramount. In mission-oriented research, on the other hand, the research problems are determined by nonscientific missions and often financed by an agency or customer with some responsibility in the mission area.

During the 1960s the international exchange of ideas and experiences in the science policy field increased considerably. The Organization for Economic Cooperation and Development (OECD) has had a central role in this development.[2] The OECD has many of the most highly industrialized countries as its members. One of its tasks is to promote economic growth and employment and a rising standard of living in member countries. This is why the OECD has devoted considerable attention to the causes of economic development.

Economic research indicated that possibly less than 50 percent of the production increase during the last decades was caused by increases in capital investments and manpower. The remaining part, the so-called residual factor, was attributed to improved education as well as to research and development (Denison 1962). R&D thus came to be seen as an important contributor to economic growth, and science policy as an important field for the OECD's cooperative efforts.

At the beginning of the 1960s, OECD activities were geared more to the quantitative aspects of science policy, more to the total amount of R&D activity and less to its content, its administration, and its relation to other policy objectives. Much attention was devoted to the collection of statistical data on R&D costs for international comparisons. Gradually, however, there was a growing awareness of the importance of the qualitative aspects of the science policy problem, about the way research should be linked to the needs of society, and how the societal direction system for the R&D area should be designed.

An important milestone in the development of OECD's conceptualization of the science policy problem was the report of the Brooks' committee, *Science, Growth, and Society* (OECD 1971). This report uses the expression *science policy* to mean "policies for the natural sciences, the social sciences, and technology." It presents the science policy problem as one of balancing two models—one centralized and one pluralistic. The models are described in the following way:

> In the pure form of the pluralistic approach, resources are assigned to each policy sector as a whole—e.g., defence, health, agriculture, transportation, housing, social welfare—and the appropriate level of R&D for each sector is then determined in competition with capital investments and service expenditures in the same field. (p. 66)

> In the pure form of the centralized model, all resources for publicly supported research and development would be given to a single agency, which would then determine the allocation of these resources for scientific and technological activity serving the purposes of the various sectors as well as for the maintenance of the health and vitality of the scientific establishment as a whole. (p. 67)

An advantage of the pluralistic or sectoral approach is that research and development activities tend to be well matched to the operational needs of social goals and more responsive to these needs. A disadvantage of the sectoral model is "that it places the long-term and short-term needs of the sector in direct competition with each other. As a result, the longer-term needs may suffer, especially in periods in which the total resources available to the sector are restricted or declining."

The Brooks report includes conclusions and recommendations to OECD's member countries. It advocates the sectoral approach in fields in which social needs suggest the requirement for a high level of research and innovative activity, but it also stresses the need for a blend of centralized and sectoral approaches. The following quotation can be taken as a summary:

> We recommend that the Member countries, while encouraging the development of sectoral policies and their interactions, should ensure the central availability of a proportion of their research funds for longer-term investigations and studies, which will probe beyond the research programmes planned to meet well-formulated problems of departments and agencies responsible for particular sectors.

In Sweden toward the end of the sixties there was a clear development toward thinking of mission-oriented research in terms of sectoral research. Symptomatic of both this development and the Brooks blend of the two science policy models was the almost simultaneous appointment in 1969 of two committees. One was to investigate the need for coordinating technological research in various areas, including defense.[3] The formulation of its task can be seen as an expression of the centralized model. The other committee was to examine the organization and future direction of research for defense. Its report, entitled *Research for the Defense Sector* (SOU 1970:54) was a clear expression of the sectoral research model.[4] While some of the main proposals of the latter committee were accepted and carried through, the proposals of the former did not result in any significant changes.

The fact that the sectoral research policy model was first applied most fully to the defense sector in Sweden cannot be considered a coincidence. Two contributing factors are clearly discernible. One is the Swedish security policy of nonalliances. This makes it important for Sweden not to be completely dependent on the import of modern weapon systems but to have the R&D capacity to develop weapon systems of its own. If the centralized science policy model were to be applied, the allocation of resources to R&D in defense would be a result of some general science policy, which would then indirectly control or circumscribe Swedish security policy. Obviously it was more acceptable to consider R&D for defense to be an internal defense problem, thus facilitating an adaptation of its R&D policy to Sweden's security policy. Another factor adding to the suitability of the sectoral research model for defense is that the defense sector is clearly distinguishable from other governmental areas. In other fields the specific sector to which various R&D efforts are potentially applicable cannot be distinguished as clearly. Also sector boundaries are often rather blurred.

The committee report *Research for the Defense Sector* introduced some new concepts and structures for sectoral science policy discussions, which seem to have influenced science policy developments in other sectors in Sweden. Two concepts, namely program responsibility and production responsibility, were introduced. By program responsibility for research was meant the responsibility for identifying research needs and for commissioning research. An agency or a committee may have a program responsibility without bearing any direct responsibility for the production of research results—that is, without having production responsibility. At that time a frequent weakness of the science policies of other sectors was the existence of problem areas for which no authority bore the program responsibility. Another fairly general problem was whether the same organizational bodies should bear program and production responsibility for research. These types of problems could now be more easily discussed.

Other features of the defense research study have also had an influence on sectoral science policies outside defense. Of particular importance is the way that the study emphasizes the link between research and long-range defense planning. R&D results are described as inputs into the planning process. Quite elaborate descriptions are given of how various types of R&D results are used in different parts of the planning process. It also pointed out that there is a relationship or information flow in the other direction as well—from the long-range prospective plans to R&D planning. However, no details about this relationship are given. It is just briefly stated that the prospective plan has implications for the need for R&D and that these should be analyzed (SOU 1970:54, pp. 65, 109).

In the earlier OECD science policy discussion sectoral planning was not mentioned, although the Brook's report referred to the need for sectoral science policies that could couple R&D to various social goals (OECD 1971, p. 66). This connection to social goals is further stressed in a more recent report, *Science and Technology in the Management of Complex Problems* (OECD 1976), as is illustrated by the following quotations:

> The reformulation of social or political concerns in terms amenable to research, the translation and integration of research results into policy conclusions or proposals have become, for example, specialised activities, leading to the development of professions or institutions which act as intermediaries between the parties concerned... (p. 11)

> Over the short-term, all sectors of policy could benefit by calling upon science to assist in the formulation of goals, and in the implementation of programmes. (p. 11)

In this 1976 OECD report the need for long-range prospective planning is also mentioned:

> Policy decisions, whether taken by the public or private sector, become increasingly irreversible because of the vast human and material resources involved and the number of people directly or indirectly affected by them. This points to the need for a means of determining the compatibility of options before they are adopted and of anticipating problems which may result. Long-term prospective planning might be one such tool. (p. 9)

In summary, this review of the historical development and conceptualization of the sectoral science policy problem has reflected not only a development from applied research to mission-oriented and sectoral R&D but also a growing interest in the linkage between sectoral R&D and sectoral goals and planning.

Strategic Planning for Transportation

The Transport Research Delegation is a research council set up in 1971 with program responsibility in the transportation sector. The way its tasks were defined shows some influence from the defense research study (SOU 1970:54). TRD was given the responsibility for planning, coordinating, initiating, and funding R&D concerning transportation—and this in areas that are not the responsibility of other governmental agencies. Other agencies with some program responsibility for transportation are the Road Transportation Agency and the National Railways. TRD's research responsibilities concern problems that cut across different modes of traveling and problems connected with transportation policies on the national level. TRD can thus be regarded as having responsibility for the R&D problems of the Ministry of Transport.

During its first few years most of TRD's research funds were used for short-term, operational problems. This can be seen as an example of the problem mentioned in the 1971 OECD report with regard to the sectoral research model. A weakness of this model was said to be a tendency to devote too little effort to the somewhat longer term problems. However, this problem was recognized by TRD, and the transport research study, which the delegation then initiated, was envisaged as a remedy. The study was called a prospective plan. I shall henceforth refer to it as the Trp study. Its task was to provide a basis for orienting R&D.

The term *prospective plan* is used in Swedish to denote a kind of long-term strategic plan that outlines alternative futures and alternative development possibilities. It can therefore be considered a kind of futures study if we use the Shani framework for distinguishing between futures studies and conventional planning (Shani 1974). Thus the basic hypothesis behind the Trp study was that it is possible to get a better overview of the need for research and development efforts in a sector such as transportation from futures studies or from some kind of long-range strategic planning of the development of the sector itself.

The general approach used in the Trp study was to make a preliminary study on the basis of a conceptual model of strategic planning and then to revise the model and apply it.

The term strategic planning was originally introduced to designate the preparation of the strategic decisions of a business firm (Anthony 1965). Nowadays it is sometimes also used in the public sector. Different models and theories about strategic planning have been developed. A simplified outline of Ansoff's model is shown in figure 11-1.

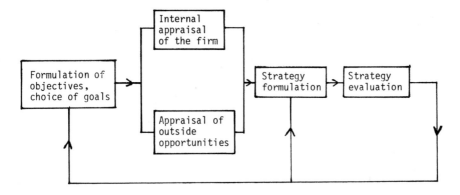

Source: I. Ansoff. *Corporate Strategy*. New York: McGraw-Hill, 1965.
Figure 11-1. Simplified Outline of Ansoff's Model

To relate planning models from different fields it is convenient to use some planning terminology. The *planning object* is the system or activity area, the development of which is to be planned. Thus in Ansoff's model the planning object is the firm. In the Trp study the planning object is the transportation sector, which has been denoted the transport system. It is taken to include both goods and passenger traffic and individual as well as public transport systems. The ways in which a planning object can be developed toward desired objectives depend on the *environment* of the planning object. The environment of the transport system consists of the aspects of society that affect the future development of transport demand and transport requirements. Examples of such aspects are population growth, working hours, urban structure, regional development, the production system, and exports and imports. On the basis of some models of strategic planning and the prospective planning model for defense the model in figure 11-1 was chosen and applied in the preliminary study of the Trp project. On the basis of experience from the preliminary study, the model was revised and the version shown in figure 11-2 evolved. The model was considered primarily a learning model, that is, a logic for collecting and structuring information so as to obtain the perspective and knowledge needed for an overview of problems and R&D needs. The model had seven parts.

Part 1 of the model is an analysis of the transport system as a system. It deals with questions such as what is included in the system and what forms its environment. It also analyzes the goals or objectives of the system. Many transport studies do not deal with goals explicitly but make implicit assumptions. It has been pointed out, for instance, that a frequent implicit goal is "the greatest mobility for the greatest number" (Adams 1977, p. 24). If this is used as a criterion in transport planning, it may cause serious problems. Large economic investments in transportation to increase mobility may mean that the contribution to an aspect of welfare is smaller than it might have been if the same economic resources had been used in other areas. Other problems connected

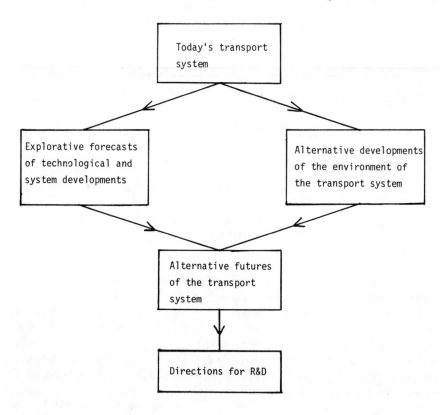

Figure 11-2. Planning Model in the Preliminary Study of the Trp Project

with the mobility criterion are that it does not take regional effects or negative environmental effects into account.

Satisfying the demand or the need for transports is sometimes taken as a goal. However, in planning we need to know future demands and future needs. These depend on future structures for production and services, which cannot be easily forecasted and which depend on societal goals. So the goals of transportation need to be related to the goals of society.

In the Trp study the approach used is to identify goals of importance for transport decision making by examining the direct and indirect effects of transports and the way they affect society. In this way five goals are derived.

Acceptable daily programs

Better balanced regional development

Efficiency of production processes

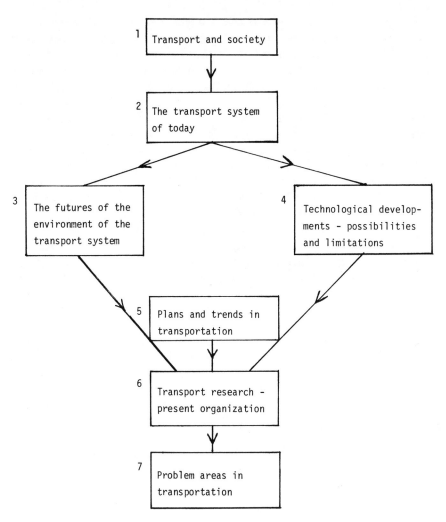

Figure 11-3. Revised Planning Model in the Trp Project

Limitation of negative side effects of transportation

Limitation of energy use

By acceptable daily programs is meant that the transport facilities should allow every individual access from his residence to places of work and to shopping and public services, with reasonable freedom of choice among action sequences.

In part 2 the functioning of today's transport system in Sweden is described and analyzed. The five goals provide the structure for the analysis. Insufficient goal fulfillment is used as a point of departure for formulation of problems. One of the problems of today's transport system is unsatisfactory daily programs in some geographical areas for certain categories of people, for instance households without a car in sparsely populated areas. Another problem is that the transport system tends to increase rather than counteract existing regional imbalances. However, perhaps the most serious problem in the field of transportation today consists of negative side effects, particularly road accidents and pollution. Another type of problem concerns the use of energy. The dependence on petroleum-based energy resources in particular must be reduced within a decade or two. The problem already exists as many of the means for reducing the use of energy or petrol have long lead times.

There are also transport problems connected with the way transports are financed or with the decision-making structure in the transport sector. Basically the aim should be to partition transports among the modes of traffic in a way that is economic to society. However, this aim raises a number of problems originating from differences in the cost structures, financing systems, and control of different modes of traffic. Other problems, again, are connected with the complicated decision-making structure, for example, the division of responsibilities between central and local authorities. In the Trp study the term *problem* is used for real-world problems. Real-world problems are not the same as research problems,[5] but if the real-world problems are considered important, they should be further examined with a view to clarifying whether more knowledge is needed to handle them and whether R&D can help in such cases.

The problems discussed in part 2 are the problems of today's transport system. These are not necessarily identical with those of tomorrow. New problems may arise. The problems of today may change in character, and the means and conditions for their solution may also change. Parts 3 and 4 of the Trp study survey developments that may lead to these kinds of change.

Part 3 is a survey of the future development of the environment of the transport system—aspects of society that affect the transport sector. Among such aspects are future changes in values. Alternative future developments of different environmental factors are outlined. By combining various factors three different scenarios are constructed, each representing a future Sweden or, more exactly, an example of a development of the environment of the Swedish transport system.

Scenario A is a kind of reference case outlining a fairly surprise-free future. In several respects it follows fairly closely earlier trends. The decrease in weekly working hours is no faster than during the fifties and the sixties, resulting in 7 or 6.5 working hours a day (five days a week) toward the end of the century. Average yearly growth in GNP is about 3 percent, the same as the growth of consumption. Scenario B represents a development with a more marked decrease

Linking Research Planning to Sectoral Planning

in weekly working hours and a slower rate of economic growth. Scenario C is characterized by energy supply problems, an increase in energy prices, and changing terms of trade. This results in the same growth in GNP as in scenario A but combined with a much lower increase in consumption. The traffic flow can be expected to increase by 50 percent to 100 percent before the year 2000 in scenarios B and C, and still more in scenario A.

Part 4 examines technological developments of possible importance for the transport system. For instance, new types of engines and fuels are examined with regard to their ability to diminish pollution and decrease energy consumption. The presentation of development possibilities and of factors that may influence developments can be seen as an instance of explorative technological forecasting.

In part 5 existing trends and plans for transports by road, rail, sea, and air are examined. One of the purposes of this part is to determine the extent to which transport problems are already being dealt with by the transport agencies as part of their planning activities. Remaining problems may need special R&D efforts. One of the conclusions is that although a very advanced planning system has been developed for road investments, there is a risk that effects that are difficult to quantify or to measure in monetary terms are not being given sufficient weight. Examples of such effects are negative side effects (accidents, emissions) and regional development effects.

Part 6 discusses transport research as an example of sectoral research. The organization of transport research in Sweden and the role of the Transport Research Delegation is described.

Part 7 deals with the research planning process and ways of categorizing R&D projects to facilitate the dialogue between the researcher and the users of research results. It can be useful in this context to structure the R&D requirements along several dimensions, for instance real-world problem area, research purpose, user of research, and research area. In part 7 the *problem area* is taken as the basis for categorization. For the goals defined in part 1, problems are discussed that can be considered to come under the heading of insufficient goal fulfillment. The choice of policy instrument for reducing energy consumption can be mentioned as an example of a problem of this type. More knowledge is needed about the design and effect of different policy instruments: Should special agreements be made with the car industry? Should new rules be introduced governing the importation of motor vehicles? Should the tax structure be differentiated so that motor vehicles with high fuel consumption are subject to higher tax?

Another type of problem area is related to the organization of the transportation sector and the principles for pricing and financing transport services. A number of problems in this area are discussed in part 7.

The Trp study thus leads to the formulation of a number of real-world problems but not directly to the definition of research problems or projects.

This brings us back to the question with which we started: What is the relationship between long-range planning and research planning?

From Long-Range Planning to Research Planning

When the need for a link between research and long-range sectoral planning is discussed, it is important to realize that this link has two different sides and thus involves two issues: the usefulness of research for planning and the usefulness of planning for the orienting of research.

Let us first discuss the usefulness of research for planning and, in doing so, compare the defense sector with the transportation sector. In defense, research results have been described as inputs into the planning process (SOU 1970). Now defense is a technology-intensive sector; most defense research aims at technical improvements in systems and system components. However, new elements cannot be introduced unless the change has been planned for a number of years ahead, because development, production, acquisition, and coordination with other parts of the system result in long lead times. This means that results from defense research actually have to be used, directly or indirectly, in the planning process; otherwise the research results will not be used at all. Most defense research is used very indirectly in the planning process. New knowledge, produced by research, often has to be integrated with existing knowledge in the defense administration and further processed before it can be used as an input into planning.

There is also some defense research that is not technology oriented but is more general in character. However, the purpose of this research is also to influence the development of defense, and thus it needs to be used, directly or indirectly, as an input into some type of defense planning.

If we now compare the use of research for defense planning and the use of research results in the Trp study, we find that there are both similarities and differences. Some results from transport research proved very useful to the study, for instance some results from transport economics and demographic and economic geography. Part of the work of the study was devoted to integrating results from such varied areas. However, most of what has been done in transport research during recent years was not used in the Trp study. This does not mean that this research has not been useful, but it has been directed toward the needs of various agencies and not intended for use in a study of the Trp type.

So if we now examine the hypothesis that mission-oriented research is useful for planning, it seems rather obvious that this is the case in areas where planning is needed. However, the research results may be used in different types of planning and sometimes in a very indirect way.

The next question to be discussed is whether planning was useful for orienting research. This is a different and more complicated matter. In a technology-intensive area in which long-range plans are used, there is obviously a need for an information flow from the planning process to research planning. It is not immediately obvious how this should be organized. In Swedish defense several methods are used in combination. One is to have researchers take part in various planning studies and, in doing so, to identify new problems together with the customers (Jennergren 1977, pp. 18-20). Another is to set up a dialogue between the military planners and the research planners. Because most defense research in Sweden is carried out at a defense research institute with permanently employed researchers, it has been possible to formalize such dialogues and to introduce a system of long-range research plans.

If we now compare the Trp study and the transportation sector, we find that there are some obvious differences. TRD sponsors mainly problem-oriented rather than product-oriented research in the field of transportation. Product-oriented technical research is funded by the transport agencies and also by a technical research council (STU). Most of the researchers financed by TRD work only temporarily on TRD problems. Thus there is not the same opportunity for developing research plans as there is in defense.

Because of lack of earlier paradigms it was not clear at the start of the Trp study how it could be used for orienting research. Any future-oriented study could of course be expected to improve our capacity to anticipate problems. One thing that this project seems to have shown is that a study of this type can help to identify and formulate problems and that it can provide material that helps researchers to formulate their research problems. However, it does not automatically result in research plans. Special attention must therefore be devoted to implementation problems.

In mission-oriented research there always seems to be a need for a dialogue between those who know the real-world problems and projects (administrators, planners, policymakers) and the researchers who know the limits of present knowledge and the way that R&D can supplement that knowledge. This need naturally persists even if strategic planning is introduced. Planning may improve our capacity to anticipate problems and to formulate them adequately, but the researchers themselves are needed in the process of translating real-world problems into research problems. During the Trp study some conferences were organized with researchers from different fields. It can be expected that further contacts would be very useful to facilitate the implementation of the study.

Futures studies or strategic planning can be carried out in many different ways, depending on the purpose of the study. A tentative conclusion from the Trp study is that if sectoral strategic planning is to contribute to the improvement of R&D planning, very specific demands must be made on the design of the strategic planning process. Furthermore, it is important to observe that there are considerable differences between different governmental sectors, and this must be taken into account in all efforts aimed at the orienting of sectoral R&D planning.

Notes

1. Several Swedish studies on these developments have been published recently, for example, Stevrin (1978), Wittrock (1978). A very complete description of Swedish science policies at present can be found in the committee report "Science Policy" (Forskningspolitik, SOU 1977:52).
2. For a review of science policy and the OECD, see Wittrock (1978).
3. Its report was *Coordinated Technological Research* (DsI 1970:9).
4. The existing defense R&D organization and resource allocation policy were fairly close to the sectoral research model. In other words, the proposed changes were not very far-reaching.
5. Wittrock has made a distinction between different types of problems which seems similar to the one made here. He defines *p*-problems as problems of planning and policymaking and *k*-problems as problems of knowledge (Wittrock 1977).

References

Adams, J.G.U. 1977. The national health. In *Environment and Planning A* 9:23-33.
Ansoff, I. 1965. *Corporate strategy*. New York: McGraw-Hill.
Anthony, R.N. 1965. *Planning and control systems: a framework for analysis*. Boston: Graduate School of Business Administration, Harvard University.
Denison, Edward F. 1962. *Education, economic growth, and gaps in information. Journal of Political Economy* (supplement) 70.
Jennergren, C.G. 1978. The planning division of FOA, its development and role. In *Trends in Planning*, ed. C.G. Jennergren, S. Schwarz, O. Alvfeldt. New York: Wiley.
Ministry of Defense. 1970. *Research for the defense sector*. SOU. 1970:54, Stockholm.
Ministry of Education. 1977. *Science Policy*. SOU. 1977:52, Stockholm.
Ministry of Industry. 1972. *Coordinated technological research*. DsI. 1970:2, Stockholm.
OECD. 1971. *Science, growth and society*. Report of the Brook's Committee. Paris.
OECD. 1976. *Science and technology in the management of complex problems*. Paris.
Schwarz, B., and Svensson, J.E. 1978. *Transportation and transport research: future directions* (Transporter och transport-forskning; ett framtidsperspektiv). Ekonomiska Forskningsinstitutet, Stockholm.

Shani, M. 1974. Futures studies versus planning. *Omega* 2:635-649.
Stevrin. P. 1978. *Mission-oriented research for societal needs* (Den samhällsstyrda forskningen). Stockholm: Liber Förlag.
Wittrock, B. 1977. *Sweden's secretariat: programmes and policies. Futures* August. pp. 351-358.

———., 1978. *Science, forecasting, and policy-making,* part 1. Report No. 4, Group for the study of Higher Education and Research Policy, University of Stockholm, S–106 91, Stockholm.

12 Remarks on the Formulation of Technology Strategy

Karol I. Pelc

The importance of technology strategy is obvious at all management levels where long-range decisions have to be made both with respect to investment and to research and development. That is why the technology strategy has to be considered at the national level as well as at the level of specific corporations and producers. One should emphasize that the difference of management at different levels corresponds not only to aggregation of specific actions undertaken in varying scales but also to various spheres of activity affected by selected technological strategies. It means that at particular levels there appear some additional factors that should be analyzed when a strategy is defined. For example, in the case of technological strategy at the national level, a strategy has implications for education policy, employment programs, or reduction of unemployment. In the case of corporate strategy, it is reflected in complementary solutions concerning contracts with partners for raw material supplies and for exports in selected markets.

The purpose of this chapter is to discuss those factors that are essential for R&D management, with particular emphasis on their role in the formulation of research policy at the national level in developing countries.

Relation betwen R&D Programs and Technology Strategy

It is commonly acknowledged that the future state of technology relies on the current state of the art of research and development. On the other hand, though, selection of R&D projects and involvement of financial resources depends on current technological strategy. A number of authors [1, 2] provide indications about the form of this interdependence, in determining the time horizons and limits imposed on particular projects given specific R&D strategy at definite stages of technological development. Recommendations concerning the role of R&D systems in the realization of technological strategy have been addressed to developing countries by the University of Sussex Science Policy Group [3].

1. The main objective of development at least in its first and fundamental phase, has to be the satisfaction of basic needs such as food, shelter, health and education, which are essential for any human being to be wholly incorporated into its culture.

2. The development of every country or region will have to be based as much as possible on its own resources—natural as well as human.
3. New technologies should not be socially disruptive so as to allow a smooth, continuous transition from traditional societies to better forms of social organizations preserving the best cultural elements of the old order. One essential prerequisite to this purpose is to provide socially useful employment to the active population.
4. The rational management of physical environment should be one of the great lines of economic and social development. In other words, it is essential to build a society intrinsically compatible with its environment.

It is not my intention to oppose these objectives. The goals are very positive. Nevertheless, they seem to be much too general to be dealt with in practical terms. Their interpretation could lead to formulation of various research programs. An intermediate solution would be to specify the conditions to be met by required technologies within a given time horizon. These conditions would already constitute an element of technology strategy.

Formulation of Technology Strategy: A Structural Approach

Technology strategy (TS) is considered here to be a group of decision rules that are used to select technologies or technological systems, adequate for desired long-term objectives of a given economic system in the light of major factors influencing situations [4].

Technology strategy has been analyzed by several authors from the viewpoint of the corporation as a part of overall corporate strategy or as a result of economic strategy [5, 6]. National concern about technology strategy has been developed in the socialist countries as an indispensable element of the central planning system and in other countries in connection with newly developing areas of technology assessment. Nevertheless, formulation of TS presents a problem of synthesis in establishing a framework for creating a coherent technology mix that fits the overall strategy of development, both social and economic.

Several types of technology strategies are described in the literature [1, 2]. Most significant are the offensive strategy, the interstitial strategy, the defensive strategy, and the absorptive strategy.

This specification has been based on practical observation. It is, however, inconsistent to a certain degree and is not characterized by structural order. The division has been made according to one specific criterion, *competitive behavior*. In order to fully explore the structure of the problem of synthesis, two other criteria are suggested: range of implementation and rate of technological change. As a result, each strategy can be determined in a three-dimensional way.

The *range-of-implementation* criterion is based on a sectoral division of technologies, corresponding roughly to division of products into various groups. Technology as a whole is considered to be a sum of specific, partly overlapping sectors. (Such a concept of technology is discussed by G.M. Dobrov and called a "family of technologies" [7].) This approach allows for a distinction of leading sectors whose development initiates and stimulates progress made in other sectors. The complexity of interrelations between members of a technological family can have a positive impact on its development, depending on the choice of the leading technology. The same phenomenon can also exert a repressive influence, resulting in the creation of an implementation barrier due to the need to attain certain minimum levels in all sectors, enabling the introduction of the required change in every single technology. Because of the nature of mutual intersectoral relations, it is necessary for the change to be made simultaneously in various sectors. The minimum level of technology required for introduction of change or for substitution of a given technology may be called an *implementation threshold*. On the basis of this criterion, one can distinguish the following types of technology strategy: single leading sector strategy and multiple leading sector strategy.

The former can be defined as a narrow-range implementation strategy, the latter as a wide-range implementation strategy. In the latter one can distinguish various types consisting of two, three, or more leading sectors. Moreover, various combinations between leading sectors are possible, such as between electronics and food production technologies, or between aeronautics and plastics. Types of these combinations are essential in establishing the priorities for research and investment, since the highest priority should be obtained by projects promoting the development of both selected sectors at the same time.

Rate of technological change and life duration of technological generations may be treated as another criterion for definition of strategy because of the possibility of varying paces of technological change. Historically speaking, the time needed for technology substitution has been decreasing. According to data available [7], it has shortened from twenty to thirty years in the 1930s to six to seven years in the 1970s. Substitution time varies in different countries; for developed countries it is approximately three to eight years. The time of various phases of R&D and implementation varies with the technological sector. An example of average times comes from the data collected by Maier [8] and presented in table 12-1.

In each case we have to deal with wide dispersion of time points: standard deviation constitutes 0.3-0.5 of the median. It is possible to control the change rate of this process, which corresponds to strategies that can be defined as maximum change rate strategy, medium change rate strategy, and moderate change rate strategy.

The choice of one of these strategy types results in intensification of R&D efforts or in a slower rate of development of R&D activity that can be more effective after all.

This aspect of technological strategy is essential for producers or countries

Table 12-1
Average Time Horizons for Various Technical Activities

	Time of the Phase in Years			
Industry	Applied Research	Technical[a] Development	Start of Production	Total Process (in Years)
Electro-industry	2.9±1.4	2.9±1.4	1.9±1.1	7.7±2.2
Production of tool-making machines	1.9±0.9	1.9±1.3	2.3±1.2	6.1±1.8
Machine production	1.9±0.9	1.9±1.2	1.6±0.9	5.4±1.7

Source: H. Maier. "Technology Assessment for International and National Technology Transfer." In G.M. Dobrov, R.H. Randolph, and W.D. Rauch, eds., *Systems Assessments of New Technology: International Perspectives.* Laxenburg, Austria: IASA, 1978, CP-78-8, pp. 61-65.

[a]Production and testing from prototypes, final production, and governmental examination.

that are retarded in their technological development and whose technology strategy can either promote technological acceleration within the existing technological generation in order to diminish the technological gap with respect to leading producers in a given branch, or take a big step forward and join the next technological generation. In the latter case, that strategy of pursuit may account for a temporary technology lag in the production sphere, for extension of the life of older technological generations, and simultaneously for intensive development of basic and applied research aiming at the preparation of a new technological generation in the same or an even shorter time than the leading producer. This strategy can also assume that a retarded producer would not repeat the whole evolution in the same way as leading producers and can allow for toleration of a technology gap for two or more future technological generations. The latter solution presents considerable risk for a delayed producer, since it requires large financial resources and preparation of appropriate infrastructure for the future technologies. Moreover, the choice of a given technology may prove erroneous. Thus it follows that the technology strategy in a given branch should determine the desired rate of change within the framework of specific technological generations and the desired time duration of generations.

Determinant Factors of Technology Strategy

Selection of a given technology strategy depends on the objective of one's inquiry: whether it is concerned with an enterprise, corporation, national economy, or the like. It seems that this situation can be defined by means of a list of attributes falling into two main categories, namely, policy-independent

Formulation of Technology Strategy

and policy-dependent factors. Prior to the formulation of a technology strategy, both these categories should be analyzed with respect to present and future conditions. It follows from substitution periods that the time horizon for a technology strategy and the estimation of attributes characterizing a given situation should embrace periods ranging from about five to twenty years. Table 12-2 shows main groups of such attributes. Each includes a set of measurable parameters that should be further analyzed. In the case of a simplified approach, one can also use qualitative estimates and comparisons of situations by introducing two or three critical levels. Configuration of such estimates indicates a feasibility or a need for some kind of strategy. It also enables localization of a given object or producer in a definite class.

It is clear that this description of technology strategy and the definition of situations determining the choice of strategy may be effective only to the degree to which specific dimensions and factors can be expressed by means of *measurable parameters*. Much effort has been made with respect to measurement of technological change. This has led to concepts such as the SOA factor at the micro level, and the technology state vector proposed by Pherson [9] at the macro level. There are also advanced studies on measurement of technological change based on dimensional analysis, published in recent years by Sahal [10]. On the other hand, there are several approaches to scaling the importance of various attributes of the technoeconomic situation such as those presented by Teitel [11]. It seems that further research in this domain should be linked with the problem of formulation of technological strategy. For the purpose of managing technological change, it seems most important to define such measures of technological change that could be directly used as components of specific strategies.

Table 12-2
Attributes Affecting the Time Horizon for a Technology Strategy

Groups of Attributes X_k *(Policy Independent)*	*Groups of Attributes X_1* *(Policy Dependent)*
Raw materials	Economic and financial situation
Minerals	Intellectual potential
Other natural resources	Scientific staff
Land	Information Systems
Energy	Patents, know-how
Environmental conditions	Educational and cultural level
Climate	Technological infrastructure
Water	Managerial experience
Soil	Social structure
Geography	Market needs or access to market market size
Manpower, human resources, demographic data	

Conclusion

Formulation of technology strategy is a problem of synthesis. The synthesis should take into account not only one criterion of relative competitive behavior but also the criteria concerning absolute measures—the range of implementation and the rate of technological change.

The first criterion permits us to distinguish between offensive, defensive, and intermediate strategies, while the two latter criteria enable the formulation of strategy by means of measurable parameters, which is essential when technological strategy is to be used as a framework for choosing R&D and investment projects.

Determinant factors for selection of technology strategy can be divided into two major groups, one of them embracing policy-independent attributes, the other embracing policy-dependent attributes. A close relationship between measurement techniques of technological change and formulation of technology strategy may be observed. On the basis of the assessment of the state of the art of this domain, a postulate may be put forward about the need to define such measures of technological change that could be applied to the historical analysis of technology development and would also be useful for transmission of directives and strategic decisions.

References

[1] Twiss, B.C. *Managing technological innovation.* London: Longman, 1975.
[2] Taylor, B., and Sparks, J.R. *Corporate strategy and planning.* London: Heimeman, 1977.
[3] Encel, S., Marstrand, P.K., and Page, W., ed. *The art of anticipation.* Science Policy Research Unit; University of Sussex, London, 1975, p. 242.
[4] Pelc, K.I. Technology strategy: determinant factors. Paper presented at the Third International Conference on Management of Research, Development, and Education, September 21-24, 1978, Wroclaw, Poland. *FRC Bulletin* 139:60-71.
[5] Ansoff, H.I., and Stewart, J.M. Strategies for a technology based business. *Harvard Business Review,* Nov.-Dec. 1967.
[6] Roberts, E.B. Technology strategy for the medium-size company, *Research Management* 19 (1976):29-32.
[7] Dobrov, G.M. A strategy of organized technology. In G.M. Dobrov, R.H. Randolph, W.D. Rauch, ed., *Systems assessment of new technology: international perspectives.* Laxenburg, Austria: IASA, 1978, CP-78-8, pp. 13-30.
[8] Maier, H. *Technology assessment for international and national technology transfer.* In G.M. Dobrov, R.H. Randolph, and W.D. Rauch, eds., Systems

assessments of new technology: international perspectives. Laxenburg, Austria: IASA, 1978, CP-78-8, pp. 61-65.
[9] Pherson, M. Modelling technological change. In *Systems analysis applications to complex programmes.* Oxford: IFAC, 1978, pp. 69-76.
[10] Sahal, D. A theory of measurement of technological change. *International Journal of Systems Science* 8 (1977):671-682.
[11] Teitel, S. On the concept of appropriate technology for less industrialized countries. *Technological Forecasting and Social Change* 11 (1978):349-369.

13 Scientists on Technology: Magic and English-Language "Industrispeak"

Michael Fores

There is a substantial tradition of scholarly writing, much of it in English, that seeks to relate two factors, "science" and "technology," to the growth and prosperity of nations. In the years after World War II this matter has often been seen, too, in terms of R&D spending, a variable that was not often considered beforehand. Yet despite the importance of the topic, there is evidence that those who have dealt with it (including natural scientists) have done so in an unimpressive and unreliable way.

Science, in English, is an indeterminate, variable entity, hovering between a number of meanings. Technology is a subject more favored for discussion by those who have never performed its functions than by those who have; it, too, is poorly defined in the literature. But above all, technology seems to be the scientist's construction and plaything; it is a piece of hysteria and deception, part of a metaworld for people to talk of rather than to observe or to live in. As such it is a part of a courtly dialect that can be called "Industrispeak" to which the scientist contributes more as a modern magician than as a man of reason.

This chapter examines some of the eccentricities that have crept into the outsider's discussion of the manufacturing process through the fact that much of the key, recent literature has been in English and in a flawed version of English, at that. It traces, in particular, the biased view of these processes that arises from the scholars' picture of technology, which is an unobserved, variable, English-language construct. It concludes that any group that is influenced by this discussion will neglect key human skills, in a flurry of enthusiasm to talk of science.

A Question of Taxonomy

Major faults that lie at the heart of a topic, a debate, or a subject area of science are often more difficult to discern than less important, less basic faults. This is because those who deal with the topic (or the debate, or the subject area) often will do so only following the acceptance of certain codes and premises that seem to make dialogue easier, but these codes may themselves be incorrect in a basic way.

Most of the prominent discussants of the English-language idea of technology have been its detractors or natural scientists who have been out to market their own wares and to stress the utility of science. At the same time, the engineers, most people's technologists, have tended to use the word in a totally different sense from members of either of the other two groups.

Thus Chomsky, a linguist, sees technology associated with the less acceptable face of power and of capitalism. "Three times in a generation, American technology has laid waste a helpless Asian country." [p. 7] The use of this weapon has come from the actions of the political group that has fought the Cold War. There has been "dominance of a liberal technocracy (in America) who will serve the existing social order in the belief that they represent justice and humanity, fighting limited wars at home and overseas to preserve stability" [5, p. 8]. Henry, an anthropologist, ascribes to technology both a strong connection to science and an amalgam of what he seems to dislike most about modern society: "If you put together in one culture uncertainty and the scientific method, competitiveness and technical ingenuity, you get a strong new explosive compound which I shall call *technological drivenness*" [19, p. 23]. See also the conclusion to this chapter; this is the crux of a common attack on industry in English.

De Bono, a psychologist, stresses the aspects of sophisticated and mass production of technology: " A plastic telephone is technology, a metal telephone even more so, but a highly polished metal telephone might well be something else." He is honest enough to realize that his idea represents "an impression rather than a definition.... The closer you get to it the more it is not there." He is dismissive enough, too, to assert that "technologists do not need to be creative.... In technology the information itself makes the decision—one only needs to find the information and then let it get on with making the decision" [3].

Snow, a physicist by background, suggests in his famous "two cultures" lecture that technology is part of the science culture, being simply "rather easy." Or more exactly, "Technology is the branch of human experience that people can learn with predictable results" [28]. Ashby, a biologist, in *Technology and the Academics*, uses the word technology to mean those subjects taught to future engineers [1, 10]. Zuckerman, a biologist, is more enthusiastic, referring to the technologist as "a member of a profession of infinite raiment," whose key position is now recognized in "all sections of society," [30] while technology itself is "what we ourselves make of scientific knowledge" [31].

Bronowski, a mathematician by background, has even gone so far as to suggest that "the evolution of man has always been culture-driven, and that the name of the culture is technology"[4]; he is in obvious dispute with the idea behind the title of Henry's book, *Culture against Man* [19], in his belief that culture is man-made. Kuznets believes that the "epochal innovation that distinguishes the modern economic epoch is the extended application of science to problems of economic production"[21]. Layton, a historian, thinks that the modern technologist is a kind of scientist: "By the end of the 19th century, technological problems could be treated as scientific ones" [23]. Mumford, another historian, considers that in the same modern period, "the invention is a derivative product" (of the scientist's general law) [27]. Finally, it is clear that for some people, such as Bronowski, [4] technology has been with man since the first fashioning of sticks and stones. For others, like Ashby [1] and Zuckerman [30], it is something much more recent.

From these brief passages it is notable first, that none of the writers uses the word *technology* in the sense that might be predicted from looking at its derivation. The suffix "-ology" usually indicates a study of a restricted topic or a discussion of that topic, as in the words philology and biology. The "technology" of all these commentators is obviously an activity (or a set of activities) whose major output includes bulky artifacts.

Second, it is clear that each author is thinking of a set of activities different from the other ones, while none is using the most common dictionary definition of technology, which is something like "the science of the industrial arts." Engineers, in contrast, use the key word to refer to a set of craft techniques, stressing the acquired skills and know-how that is needed in manufacturing: perhaps "the arts used in making useful goods" [8].

Given this plurality of meanings of one word that is in common use in English, it is worth making these points. (1) Any discourse that is informative rather than simply decorative depends on reliability in the use of the words and symbols chosen. (2) Any individual, however practical and down-to-earth he may imagine himself to be, is still influenced by theory, generalization, and models that have been put together in an attempt to deal with complex events, which are often embedded in the language adopted. Life is so involved and unpredictable that we all need these constructions (closely defined words and models of behavior) to help plot a course through the murk.

One obvious difficulty with the treatment of *technology* in English, which is apparent from the passages quoted, is that the scholars have controlled the meaning of the word so poorly that they have broken the most elementary rule of scientific discourse: an amp is an amp and is not an ohm. In this case, we have the biologist's equivalent of Fido the dog who is mysteriously found mewing or Neddy the donkey found growing a long, elegant trunk to help him drink water from a pond. See "Science and the Cambridge aberration" [14].

A second difficulty is equally basic, stemming, too, from the scientist's forgetfulness of the rules of dialogue. The scientific process is really very simple. It involves observation followed by some form of classification and analysis of observations followed by publication of an account for general use and for testing by others. When most scientists deal with the topic of technology in accounts of their own, they are discovered to be talking about things that they have not observed. Malinowski contrasts science which is "born of experience" with magic which is "made by tradition." Magic "springs from the idea of a certain mystic, impersonal power, which is believed in by most primitive peoples"; it is also "based on man's confidence that he can dominate nature directly, if only he knows the laws which govern it magically, [and] is in this akin to science" [24].

The rest of this chapter expands on these propositions:

1. Those scientists and scholars who discuss the English-language idea of technology are normally discussing matters with which they are personally so unfamiliar that the evidence on which they base their views is what an English court would throw out as inadmissable hearsay evidence.
2. Such discussants are thus found talking, in Malinowski's sense, within a tradition rather than from experience.
3. Most of them, by opting to talk of technology at all, do so because they believe that there is a "certain mystic impersonal power" that guides the "easy," "uncreative," "derivative" manufacturing process. This is in the manner of the quotations from de Bono, Henry, Snow, Mumford, and Zuckerman.
4. This power is called science, which is itself a variable entity in English.
5. To the extent that these discussants are talking of technology expressly as scientists, they have invariably forgotten the first rule of science, so they have taken up the role of magician, able to speak up as they do with such apparent confidence only because most of their readers and listeners know as little of the manufacturing process as they do.

The Trouble with "Science"

To some extent the flaws in the well-used English-language taxonomy are enough, in themselves, to allow scientists to talk of technology and so to increase the assumed importance of science. Just by making themselves heard and respected, scientists have persuaded the layman against most of the evidence that the technical processes of manufacture are a kind of applied science. If it were not for a popular acceptance of this idea and the science-technology linkage implied by accepting it, there would be no cause for the layman to bother to listen when the scholar discusses, at length and in a seemingly informed fashion, matters of manufacturing with which he is rarely involved on a day-to-day basis. After all, no one in his right mind would go to a lawyer to ask advice about the health of his cow or to a veterinarian for advice about taking his employer to court, unless he thought that there were links between them.

I shall comment on some of the evidence on the applied science and innovation topics, both of which seem to be firmly founded in the "Industri-speak" tradition of discussing a metaworld whose characteristics are built up more by tradition than by close observation of events. (See also [9, 12, 15-18].)

Here I give a basic point about the popular grasp of the utility of formal knowledge.

The English word *science* has three core, conflicting meanings in modern usage [9, 18]: (1) *Science Mark I*, equivalent to the German *Wissenschaft,* as in the English phrase *social science*. (2) *Science Mark II*, equivalent to the German *Naturwissenschaft,* as in the common meanings of the English phrases *science sixth form, pure science*, and *basic science*. (3) *Science Mark III*, equivalent to the German *Naturwissenschaft* plus some of *Technik*, as in the English phrases *Science Research Council* and the *science culture* of the "two cultures" of Snow [28] and others.

Besides this difficulty, discussants of science tend to use the same word in a single passage to mean both a process and the product of a process. Thus the science of science is the product of the social study (Science Mark I sense) of the process of the area of activities that English speakers group together to call science in the Mark III sense of the word. It tends to be an examination, in the manner of the sociologist or the psychologist, of the activity of engineers and natural scientists.[1]

The English-language Mark III meaning of the word science can only, I believe, be given the same title as the Mark II meaning (natural science) by those discussants who accept the underlying applied science model to describe the essential features of the technical functions of the process of manufacturing. Furthermore, I believe that almost all the discussion of technology by scholars and scientists has picked up the same, incorrect, underlying model of the processes that German speakers know as *Technik* [15]. For most of these discussants an accepted basic premise is that technology is applied science.

Because of the most basic characteristic of scientific knowledge, there can be no sensible phrase applied science. Engineers and natural scientists do not react to the same set of events in a similar way, as Snow holds that they do [28]. Empirical evidence suggests that formally expressed knowledge is rarely the principal dynamic force behind technical change, in the way that Kuznets and others imply that it is. See, for instance, the theoretic proposition of Price (chapter 14) and Sahal's empirical work (chapter 9).

The strong linkage, based on assumptions about applied science, between the Mark II and Mark III ideas of science, which underlies the English-language two-cultures idea of Snow, is an important element of the constructed world of the Industrispeakers. Yet this link is insubstantial, in logic, however much discussants of technology may have chosen to have observed the matters that they discuss.

The applied science phrase is nonsensical primarily because formally expressed knowledge of events, Science Mark I, is a public good in the economist's sense of the term; see also an extended discussion of the science of science idea [9]. That is to say, scientific knowledge can be freely used or tested, by anyone who wishes to without any payment or record, on transfer from the scientist who produced it to the potential user or tester. In this sense it is like the air we breathe: there is no decrease of its stock on transfer; it is not a scarce good in the economist's sense; it can spread through the world with little hindrance.

The phrase "applied science", in this sense of a body of knowledge, implies that the matched phrase *pure science* (or perhaps "basic science") means something recognizably different from it. However, because of the characteristics of written knowledge of events, there is, in truth, no discernible boundary to be drawn between the two areas. Knowledge of events simply has the status as knowledge. This is so whether or not anyone uses it or will use it in the future in practical settings. It is so whether the producer of the knowledge alleges that he worked for love, for money, or for the good of mankind. It is so whether it costs millions of pounds to derive or whether it was produced in half an afternoon. Knowledge remains knowledge whether it is used, applied, praised, insulted, translated into Chinese, pinned on the wall, put down in books, or published in newspapers.

In the discussion of Science Mark III, however, the applied science idea normally has a second meaning, making it much the same as the popular, variable concept of technology. As with the Mark III idea itself, this means that the notion of science, which has a basic root derived from the Latin word for knowledge, has now been transformed into something that includes parts of a process that might be called "artifaction" (or the process of making useful artifacts).

One objection to calling parts of this process applied science is that (like the objection to Zuckerman's definition of technology, "what we ourselves make of scientific knowledge") the phrase is not informative in any real sense based on observation. Sculptors, cooks, detectives, and even natural scientists make objects as various as statues, soups, reports on crimes, and reports on natural events with the use of scientific knowledge, though, of course, we are never sure when this knowledge is used because of the nature of the transaction process from scientist to user. A sculptor, a cook, a detective, or a physicist could as well be described as an applied scientist as an engineer; so could a journalist or a writer of detective fiction.

A more fundamental objection to the use of the phrase applied science, however, is that it rests on a circular argument. The modern concensus of empirical study is that most technical change in manufacturing is not influenced directly by the new use of scientific knowledge (chapter 14). But even if it were influenced directly, there is always the key element of technical skill involved, which tends to differentiate one technical solution from another one, if only because scientific knowledge is freely available. That is, those with the most competent technical specialists will tend to win the day; so the quoted opinions of de Bono, Mumford, Kuznets, and Zuckerman seem to be incorrect. However, English speakers who use the Industrispeak idiom and believe in the technology-as-applied-science idea, tend to think in this way:

1. Engineers, being technologists, are also applied scientists because they learn physics and chemistry off the job. Therefore:

2. Technology is the main technical process in industry being staffed by applied scientists. Therefore:

3. Technology is part of the broader science (pure and applied, included). Therefore:

4. Engineers should learn physics and chemistry off the job. Therefore: 1, therefore 2, and so on.

As Langrish observes as the result of study, "What industry needs is not so much the new discoveries and new knowledge produced by pure science (*sic*), but the people who have absorbed the condensed accumulation of past (natural) science which can be brought to bear on the problems of industry" [22]. Many of these people named will, of course, be engineers—though not all. The fact that they may use Science Mark II knowledge in the course of pursuing their art no more makes them applied scientists than a composer of symphonies is one also simply because he has studied the science (expressed formal knowledge) of music.

Experience, Metatalk and Deskilling in Accounts

Despite the warnings expressed up to this point in the chapter about metatalk on the topic of industry, it is clear that there have been many reports of events in parts of manufacturing that are reliable and informative. My experience is that the reliable reports are normally so because their authors have kept close to observed detail; these authors have been wary, too, of making conclusions of potential general applicability. Such reports have fed very slowly into the consciousness of the macro observers whose comments on technology I quoted and who have helped to construct the Industrispeak dialect referred to in the subtitle of this chapter.

There is no opportunity to deal in great detail with Industrispeak here. Elsewhere Rey and I have attempted to show that the ideas of industrial revolution and technological revolution (Drucker [6]) are suspect; we have also suggested that much of the debate on innovation has been flawed by another basic error of science—the dramatic has been noted and assumed to be typical [16]. Sahal's work is particularly welcome as it overcomes this potential difficulty by focusing on certain measurable, functional aspects of technical change (chapter 9). However, something of the peculiarities of the use and impact of the English-language, Industrispeak dialect must be apparent.

Some people become familiar with what is really going on in manufacturing settings and aim to write accounts of what they observe. See, for example, reports on work done by Hutton and Lawrence [20] and Mant [26] with which I have been involved. Other people are happy enough to make few, if any, observations themselves. So as the result of reading the accounts of others (many of them nonobservers, too, and thus nonscientists of manufacturing), they will write of "technology," "industry," "science," "skilled workers," "practical work," "pure versus applied," and the rest, and they will do this as if they were familiar with their own chosen subject matter. It is in this way that any metaworld is constructed, in the course of discussions and remarks about subjects by those who are not personally familiar with them.

I believe that the basic model that is invariably used by those who discuss technology in English simply by entering into that discussion is quite wrong. This could be called the STH (science-technology-hardware) model of technical change, and it is uninformative in two respects: the technology element is redundant, and formal knowledge is only one input for the artificer who controls the technical process. The single result of the discussion of technology that is potentially the most disastrous is the element that neglects personal skill and artistry by accenting science; it does this by assuming incorrectly that these are interchangeable factors. To help English speakers to disentangle the confusion involved, Rey and I have suggested that the use of the German-language *Technik* idea as a useful means to understanding real events [15].

Of course, no report or scholarly treatise can manage immediately to deskill any individual, or any group of individuals, or any nation: that is to say, an accounting process cannot take away directly the exercise of skill. Scientists are in business primarily to account for what happens, not to change it. Yet the cumulative influence of written accounts that are produced in a tradition that neglects skill and artifice can have the result of changing national cultures. In the particular case in question, use of the Industrispeak dialect in English tends to neglect all those personal skills possessed by people employed in ordinary work places except those classed as manual. It helps to produce a biased accounting tradition that stresses the use of knowledge, as in Drucker's idea of "knowledge-workers" [7], or in Bell's idea of a "knowledge society" [2], or as in the covert belief in applied science of almost all the discussants of technology.

Even after observing the process of their own work, scientists are prone to neglect personal skill. A process that is really an artful trek through the unknowns of experiment is written up later as if the way had been well planned beforehand [11]. This point is quite separate from the fact that scientists (Mark I or Mark II variety) are essentially accountants who will, at best, note what they can discern in the subject area before them. It is very difficult to isolate what human skills are and what they consist of, for informed comment to be produced on the topic.

This point is separate from the fact that, in their poor-quality commentary on industry scientists and scholars have adopted the cloak of magicians. Magic has always been able to wish away the need for the exercise of skills by taking away the need to work. Mansfield, for instance, has performed the magical feat of transforming technology from knowledge into hardware in two pages (40 and 41) of his best-known textbook [25].[2]

The scientist's main aim is to produce an account of a set of events (or the properties of a material) that is as near as possible to the truth. In this way, such an account might be termed a rational statement, for the scientist's statement may be able to say something about the reasons for things. In this way, too, the scientist might call himself "a man of reason."

At the same time, even the English-language habit of calling the process of scientific inquiry and the product of the process by the same name, science, does not transform the process of inquiry into a rational undertaking. The evidence indicates that working scientists are very poorly informed of the work of the scholars of the philosophy of science or of the scientists of science. The process of inquiry into the complexity of observable events, the scientific process, is one that is invariably guided by the hunch, artistry, and the special skills of the researcher or inquirer. It is rarely informed by general statements and laws in the manner that those who talk of technology imagine that the working engineer is informed about what to do in his work by science.

In the dialect of Industrispeak and in much of the English-language discussion of science, confusion has been wrought between science-as-a-product and science-as-a-process. The label "man of reason" has been misread, with the false implication that the scientist will be able to do his work in an ordered rational way, simply because the product of his work is supposedly a rational account of real-world events [11]. This is part of what Sorge and I have called the rational fallacy. It is part of an apology for an activity rather than any observable tested account of it [17].

The slippery, variable nature of the idea of science in English is the key to realizing how it combines with the technology construct to form a major support for the dialect that I have called Industrispeak. This is a courtly dialect preferred by those who have not bothered to observe closely events they discuss with apparent familiarity. See also the mysterious disappearance of *homo faber*, who, of course, is the archetypal skilled human being contriving to make from the material world around him things that did not exist beforehand [17].

Close empirical observation of the technical process of so-called industry tends to be upsetting to the discussants of Industrispeak. It disturbs the certainty of their generalizations. It also tends to raise in importance those technical skills of the manufacturing process that the scientists of technology inevitably manage to neglect, if only because they are difficult to isolate and to discern. To discuss the technical functions of manufacturing without explicitly noting the engineer's skills is to regress to a fairy-tale world, such as that which the child believes in when he is told of the role of the stork in human birth. Indeed, the alleged role of science in the English speaker's constructed technology is very like the stork's role in the creative process.

Conclusion

There are a variety of ways to outline the key errors and confusion of the scientists and scholars, writing in English on the topic of this chapter. Here I have suggested that these commentators tend to misunderstand science itself and to take up the role of magician in an attempt to sell the importance of the output of their activity, formally expressed knowledge.

Like Price (chapter 14), I would have preferred to see these commentators going about their efforts in a different way. This is partly because I would rather see those who class themselves as scientists telling the truth rather than telling fairy tales; fiction from the novelist is usually more engaging. It seems to me that the metaworld produced by those who discuss industry in English has encouraged the English-speaking people to neglect the skills and arts of an active life in a recognizably separate "doing and making" (third) cultural area. (Industry tends to mean "any place of work except those like my own" to the contemporary English speaker).

Ultimately all the biased accounts or technology have contrived to shift the culture of English-speaking people and have influenced their perception and behavior; when magic is in the air, work is not such an attractive idea. As to the implications of what seems to be a systematic attempt to deskill almost everyone, the chance of contagion is obvious; English is the linking language for discussion of many of these topics.

Everywhere the scholars have set themselves up as priest-magicians in an irreligious age. But the English speakers amongst them seem to have been the most successful of all, producing a web of mystification over a range of human skills, which include management as well as technology. The former idea is the humanist-arts answer to the applied science notion of the other side of the two-cultures divide [13, 14, 26, 29].

I do not ask that scholars cease to concern themselves with the topics of technology, industry, and management. However, the locations of activities of this type are notorious for being places in which people have to function with only a fraction of the information that they would like to possess. There is always hurry and risk involved. The real art is to beat the next man to the market, but the particular skills enshrined in the art are difficult to describe [13]. To the extent that scholars neglect these skills by failing to recognize them or by categorizing them as being too difficult, they will continue to tell the fairy tales about manufacturing that this chapter has dealt with. They will also continue to deliver to the layman's door a set of fairy tales about scientific inquiry.

I raised the idea of courtliness because this chapter concentrates on the discussion of a set of events from afar in the manner in which a courtier looks out from a capital city at the country that he and his colleagues claim to rule. Following Henry's wording, I suggest this proposition:

> If you put together in the main councils of a national culture, uncertainty about what goes on in manufacturing and a distorted picture of the scientific method, you get a strong new explosive compound that I shall call technononsense; this leads to a decline in national competitiveness and technical ingenuity.

In the accounts by scholar-courtiers of the ordinary world of work, we can read of a stream of revolutions, crises, insights, breakthroughs, discontinuities, and the rest, composed in a manner that looks needlessly hysterical to those who are more familiar with ordinary work places. There is something more embracing than simple English-language Industrispeak at work, to produce overdramatized, unreliable accounts. I believe that the human mind is more easily drawn to tales of high drama than to measured accounts of what really goes on.

This does not mean that I condone the mythmaking of the scientists whose work is criticized in this chapter. Rather I do not believe that their attractive tales of magic will ever be classed in the correct shelves of libraries until engineers, businessmen, and others from the third culture speak up for themselves. Without such accounts, we will continue to see a worldwide tendency toward deskilling man, through belief in courtly, "scientific" accounts about technology.

Notes

1. At the risk of confusing the reader, in an attempt at clarification, I point out the following English-language equivalence: Science (Mark II sense of science) equals natural science (Mark I sense of science). In the title of this chapter, I am using science in the basic Mark I sense, which turns out to be the only reliable one to use and is the basic sense anyway.

2. Mansfield defines technology explicitly as knowledge, then argues that technological change would best be measured by a total productivity index, which would measure results of changes in hardware.

References

[1] Ashby, Eric. *Technology and the academics.* London: Macmillan, 1958.
[2] Bell, Daniel. *The coming of post-industrial society.* London: Heinemann, 1974.
[3] de Bono, Edward. *Technology today.* London: Routledge, 1971.
[4] Bronowski, Jacob. Technology and culture in Evolution. *Cambridge Review* 91 (8 May 1970).
[5] Chomsky, Noam. *American power and the new mandarins.* New York: Pantheon, 1969.
[6] Drucker, Peter. The technological revolution. *Technology and Culture* 2 (Fall 1961).
[7] Drucker, Peter. *The effective executive.* London: Heinemann, 1967.
[8] Fores, Michael. What is technology? *New Scientist* (15 June 1972).
[9] Fores, Michael. Science of science: a considerable fraud. *Higher Education Review* 9 (Summer 1977).

[10] Fores, Michael. *Scientists on technology: a confusing saga.* London: The Free University of Kensington, 1978.
[11] Fores, Michael. Stories about science. *Higher Education Review* 10 (Summer 1978).
[12] Fores, Michael. Whatever happened to *homo faber*? Paper given at British Association Annual Meeting, (Bath: September 1978).
[13] Fores, Michael, and Glover, Ian, eds. A Proper Use of Science, in *Manufacturing and Management.* London: (Her Majesty's Stationery Office 1978).
[14] Fores, Michael. Science and the Cambridge Aberration, *Cambridge Review* 100 (25 May 1979).
[15] Fores, Michael, and Rey, Lars. *Technik*: the relevance of a missing concept. *Nature* 269 (1 September 1977); and a longer version: *Higher Education Review* 2 (Spring 1979).
[16] Fores, Michael, and Rey, Lars. Technology and innovation: confusions based on error. London: Mimeographed. 1978.
[17] Fores, Michael, and Sorge, Arndt. The rational fallacy. IIM Discussion Paper, Berlin. International Institute of Management, 1978.
[18] Fores, Michael, and Sorge, Arndt. Two Cultures, or three, or one? *Encounter*, forthcoming.
[19] Henry, Jules. *Culture against man.* New York: Random House, 1963.
[20] Hutton, S.P., and Lawrence, P.A. *Production managers in Britain and Germany.* Southampton: 1978.
[21] Kuznets, Simon. *Modern economic growth.* New Haven, Conn.: Yale University Press, 1966.
[22] Langrish, J. Does industry need science? *Science Journal* (December 1969).
[23] Layton, Edwin. Mirror image Twins. *Technology and Culture* 12 (October 1971).
[24] Malinowski, Bronislaw. *Magic, science, religion, and other essays.* London: Souvenir, 1974.
[25] Mansfield, Edwin. *The economics of technological change.* New York: Norton, 1968.
[26] Mant, Alistair. Authority and task. in Fores, Michael, and Glover, Ian, eds., *Manufacturing and Management.* Her Majesty's Stationery Office 1978.
[27] Mumford, Lewis. *Technics and civilization.* London: Routledge, 1934.
[28] Snow, C.P. *The two cultures and the scientific revolution.* Cambridge: Cambridge University Press, 1959.
[29] Sorge, Arndt. The management tradition: a continental view. in Fores, Michael, and Glover, Ian, eds., *Manufacturing and Management.* London: Her Majesty's Stationery Office, 1978.
[30] Zuckerman, Sir Solly. The Image of Technology. London: Oxford University Press, 1967.
[31] Zuckerman, Sir Solly. Society and technology. *Journal of the Royal Society of Arts,* August 1969.

14
A Theoretical Basis for Input-Output Analysis of National R&D Policies

Derek de Solla Price

Two major themes have always dominated science policy theory, the study of science indicators and the economic analysis of research and development. (1) How can one define and measure the social and economic value of scientific and technological research? (This is often called the question of the social license for research.) (2) How can one define and measure the way in which investment in development (sometimes called research and development) causes output of new and valuable products? (This is sometimes termed the problem of input-output analysis of industrial R&D.)[1]

The burden of this chapter is to cut the Gordian knot in both questions by suggesting that their formulations are misguided, derived from historically false, or at least naive, views of the role of science and technology in society and of their manner of interaction. Both questions need to be rephrased, to the point of inverting their causal implications. If this is done, I maintain, one can derive not merely a logical and coherent conceptual scheme for science policy studies but even a viable means of quantitative analysis that can squeeze policy-rich results from available fiscal data.

Before the questions can be formulated we must define or, more realistically, explain the basic concepts used in them. As has been remarked many times, one of the most powerful general methods we have for the analysis of science in society is to follow, not a line of its philosophical structure, but the Mertonian line of sociological enquiry into its norms. The essence of this line is a very powerful principle that distinguishes scientific creativity from all other human creativity. Scientists act "as if" (my allusion is to the Vaihinger *als ob* philosophy) there is only the one world to discover. It may well be that scientific discoveries as perceived are nothing but a secretion of our brains and depend on the standpoints and concepts peculiar to particular cultures and societies; nevertheless, scientists feel a heavy constraint of universality. They appear to have no alternatives or rarely more than a frontline choice between a couple of modes. In a strong sense there is not, never has been, and never can be an alternative physics or an alternative chemistry. Even when we have the evidence of a different and autonomous civilization on the dissecting board of history, as we have for early China, thanks to the genius of Joseph Needham, we get a strong feeling that though the technologies are very different, such science as works bears an almost uncanny resemblance to that of any other culture.

It is this that motivates the norm of universality, and as a direct consequence, the social organization of science since the mid seventeenth century has included the unique characteristic that the act of human creation in science is incomplete without both the validation and the automatic evaluation that is accorded by an unorganized but remarkably effective concerted effort of the worldwide community of fellow scientists. Each scientist has experienced, probably quite often, that a perceived discovery may involve an erroneous piece of mathematics, a terminological fallacy, a procedural blunder. A gem of wisdom for its originator may be a worthless platitude to everybody else in the field. A dropped remark may be held universally as a stroke of genius. We can never know our skill or lack of it in communicating in an effective manner to the right audience in the right moment.

Because of this the literature is not just a mode of storing or communicating information. It is not a transmittal of discovery. It is in and of itself the end product of research in science. Like a promise that cannot be kept until it has been given, scientific discoveries can only become the property of the discoverer by the act that involves giving them away freely to the world community. It seems unjust, but if Henry Cavendish insists on not publishing his discovery of the law relating electrical voltage and current a century before Georg Friedrich Ohm, he is not the discoverer. This is no place to go into the consequences of this system in giving rise to a mechanism for resolving the typical disputes over contested priorities due to the fact that discoverers and discoveries form a Poisson distribution in their tendency to overlap in multiplicity. Nor is it the place to elaborate on the theme that painters and poets cannot have such problems, for their creativities are in terms of human individual personalities rather than in the analysis of a common universe in common terms.

The universe of openly published and available, cumulating scholarly literature in all the sciences is not particularly difficult to define from a librarian's list of scientific periodicals or, better still, from a journal citation index that generates from the use of references the entire body of connected literature and can even do so on a field-by-field basis. It follows that all the research literature in this corpus must become common property in the very act of discovery and the concomitant validation and evaluation. As a direct consequence of this, all work of research whose end product is of this form must be considered a free good in the economists' sense. This does not mean that the research is without value. It is a good; indeed, to take the traditional claim, it may be of enormous value to the world in giving wealth or might, in curing dread disease, reducing misery, and lessening the fragility of our environment. Even, it must be added, an increase in our knowledge of the world is an addition to culture, perhaps an object of wonder and aesthetic beauty as much as any other great creative work of mankind. Nevertheless, it is a free good, belonging to the world as a whole and not particularly to any one language or nation, certainly not to any particular research laboratory or industrial company. It does not even belong, in a strict sense, to the discoverer who has released it to the world in the act of discovery, and special steps must be taken to award the credit, in a moral sense, where there is no real property that can be assigned. Planck may have just as much individual style and creativity as a Beethoven or Picasso, but Planck's Constant tells no tales of the language, religion, or politics of its inceptor.

If research in the international knowledge system is a free good, that is not to say that all work involving science and technology is also free. Scientific journals cannot be free, for somebody, either a commercial or a nonprofit publisher, must be paid for their production. These concomitant expenses associated with research are relatively trivial compared with the main burden that those who also do research, or those who are trained in science and technology but never do this sort of research, may well have an output that is not free but valuable.

Any time there is an output other than the publication into an open knowledge system, it may be that this output is intrinsically valuable. Often in the academic world, people whose output is also free research have a primary job function, which is to teach. In industry the primary job may be the planning, designing, testing, and selling of a product. In the civil service and in other service professions the output may well be the dissemination of advice and nonpublic knowledge that is valued by the client or by the nation as represented by its government. Furthermore, lest we be misled by so much emphasis on new knowledge and innovation, it must be remembered that most science and technology labor is done behind the research front rather than at it. The bulk of medically trained professionals are physicians engaged in delivering health care rather than in winning new knowledge. Most engineers work in production rather than in research and development. Most people with a degree in science, even those with a doctor's degree, are using their knolwedge rather than extending it.

This activity of qualified scientists and engineers in producing intrinsically valuable services rather than the free good of knowledge is relatively easy to conceptualize and enter into a national balance sheet. It has been the particularly visible and obviously crucial role of research that has been difficult to evaluate. I now suggest that although research into the international corpus of knowledge is a free good and therefore has no intrinsic value, it certainly has a calculable extrinsic value. This derives from the fact that the world corpus of knowledge is not static, but growing in bulk at an exponential rate of about 7 percent per annum. That is a general trend from the last two centuries with remarkable little fluctuation from time to time or from land to land. Furthermore, even if there is a current saturation and slowing for the most advanced nations, this affects only slightly the world rate for the next several decades, probably longer.

The result of exponential growth of the world knowledge system is to produce an automatic obsolescence of individuals who have been trained to the research front and enabled to give some intrinsically valuable product as a result of that training. If they acquire no additional knowledge over the subsequent years, then without erasing or debasing the knowledge they already had, they will fall back 7 percent each year with respect to the newly augmented corpus of learning that has become available. In about ten years from the end of their training they will know only half of all the knowledge there is to be had within the same specifications as that of their original training. If their value depends on the level of their grasp of the field, their worth will decrease at the rate of 7 percent each year. There may well be fields of study and life-styles of people that make it worthwhile to permit this obsolescence to take place. The value of the worker's product may depend more on their basic skills, far behind the research front, for example, if they teach at an elementary level or if they are involved in making quality control tests with an unchanging method. Another possibility is that the new learning acquired by experience in the job actually fits them for new duties of a different sort as fast as the old qualifications obsolesce. For example, it is quite usual in some forms of employment of scientists and engineers for research to be gradually replaced with administrative functions. Yet another possibility of avoiding the wearing out of an initial research front qualification is for the people concerned to seek to acquire at least some of the new knowledge in their field by reading journals that make a special point of detailing fresh advances by state-of-the-art surveys or by taking refresher courses. Both procedures are by no means uncommon among physicians and some engineers.

Because professional skill at research front levels is so rare and represents a very heavy personal commitment and investment, it has become much more usual to continue research front capability by simply continuing research. It is, after all, an acid test of competence in the total knowledge of the field to be able to add to the international growth of knowledge. As it can be put aptly, the best way to tell a mind is full is by watching it run over. It is this that gives a suitable extrinsic value for engaging in research as admission to the world's knowledge in its continuous exponential expansion. Research activity, even though it produces a free good as its direct result, has the indirect result of preventing what would otherwise be a 7 percent per annum obsolescence. Since the investment that would be lost by such obsolescence is approximately four years of postgraduate training, it follows that a reasonable price to pay is 7 percent of four years. We therefore suppose that there is an extrinsic value to research that will yield continuous state-of-the-art capability in exchange for 28 percent of a person's annual work. This is consistent with the conventional wisdom that scientific teachers and industrial researchers should be able to indulge in free research for something like one-third or one-quarter of their time. It also agrees with the implicit convention of the U.S. National Science Foundation and universities that professors work a nine-month year, need a one-month vacation, can be hired for research for the remaining two-month period, and can therefore be enabled to augment their salary by 22 percent for free research.

There are, of course, considerable difficulties in admitting that research that gives a product only of knowledge is a free good and of extrinsic value only as an antiobsolescence measure in connection with an intrinsic value for some other goods or services delivered. The entire postwar entrepreneurial tradition of research proposals, as well as the plea that goes back to the days of Leonardo or even Archimedes, has been built on claims for a direct utility of the labor of scientific research. Any reasonable student of science policy must shudder at the thought that we might admit that none of this research, not even that which we brandish as applied, has value other than as a free good. Visions arise of collapse of even the scant funds that remain. One could almost guarantee that politicians would claim that research has no value whatsoever, intrinsic or extrinsic, and one could also predict their committing the cardinal sin of science policy: the making of sudden changes that always are disastrous to the merits of people and programs.

Another serious difficulty is that for various historical and practical reasons we have adopted a practice in some cases of providing full-time research careers for which there can be no extrinsic merit since there is no product other than the enrichment of the corpus of learning. A few such posts in special institutes of advanced study are easy to justify for any society that values culture for its own sake and that will gladly support a few poets and sages for their decorative value. In a developing country it might be prudent to create a reservoir of talent by offering such posts to brilliant nationals who would otherwise be lost from educational and industrial systems that could not support them to those countries that were attractive. In this manner organizations of full-time researchers were formed in the British Empire, for example, the Councils for Scientific and Industrial Research (and variants of this name) in Canada, Australia, and India. We also have the tradition of the Leibnizian academies of science that permeates the Soviet Union and the socialist countries. The academy employs scientists in a research capacity, removed from the university tradition, who are held accountable for their direct utility to the nation only with difficulty. There are also various types of research in which the workers might reasonably complain that their special field was such that it could not be worked on in any part-time capacity. Such a claim is frequently heard of research in many biomedical specialties and in those parts of modern physics requiring a staff of full-time workers who tend very large machines that are virtually institutes in themselves.

Though all of these cases seem reasonable, I suggest that we pay dearly for heeding these special pleas. The people engaged in pure research have no direct interface with activities that have immediate social value to the supporting country. Those who have the new knowledge are not providing teaching and training, and they have no direct chance to see pieces of new knowledge applied to new products and services. Undoubtedly this is why one hears bitter complaints about valuable research reports that are laid on the shelf so that no actions are taken and all the investment in research is thereby lost. In the case of biomedical research we are buying knowledge of intense interest to the whole world and paying for it by ensuring that such knowledge is unlikely to be applied directly in teaching, in industry, or in the delivery of health care in the nation most active in its promotion.

This method of analysis is in agreement with British science policy experts in the Rothschild reform of funding. Apart from the support for basic science, which our new method can justify and compute on grounds of extrinsic value alone, British policy is that if technical experts seek funding on the basis of direct utility to society, they should be placed in some contractual agreement. Whether it is a valuable product in the form of hardware or software or simply services that are to be delivered, there can be an agreement between the expert and the customer, and thus the contract provides an economic valuation. If it cannot be contracted for, one is forced to conclude that the claim for direct utility cannot be honored. That disposes of a great deal of nonsense in the search for funding for such useful objectives as a cure for cancer. One does not question the intention of researchers to be useful through their work, but the possibility of contracting for the outcome acts as a test of whether the merit is to be intrinsic or extrinsic. Perhaps it should be stated that it is not only the researcher that is so often forced by the present entrepreneurial system to somewhat questionable methods. Very often in the course of national science policies, a desired result would be costly, and the tactic is adopted of calling instead for research, the much cheaper commodity. Research into urban transportation, for example, is much cheaper than the purchase of urban transportation technology. The origin of many of the valuable research reports into practical problems may be an expensive way of reaching a required objective. A better tactic in the case of urban transportation would be to contract by bidding for whatever technologies are required. A contracting company, confident that they could do a satisfactory job, would have to include the cost of their adaptation of techniques in the cost of the contract.

The answer to the first question raised at the beginning of this chapter is therefore as follows.

1. The social license for research is to be judged first by the product that is the principal end result. If that product is knowledge entered into the international system of scientific and technical journals, the research is a free good, and its value is the extrinsic one of preventing obsolescence of the workers and giving them access to the world's knowledge system.
2. If the product of the research is other than the free good of knowledge, the value may be determined, preferably by a contractual relationship, for the delivery of goods and services.
3. Only in very limited cases of support for purely cultural objectives, humane consideration of otherwise unemployable persons, should support be given

for research leading into the international knowledge system where that research is taken to have intrinsic but uncontractable value. Full-time research careers without otherwise useful services are to be discouraged.

This clears up many problems of the social and economic system for research support, and it leaves in plainer view the other central question of the way in which scientific and technical knowledge can be applied to useful ends. We may now assume that in exchange for entrance to the world's research knowledge, a typical worker in the pharmaceutical or computer industries, for example, may devote two-thirds or three-fourths of his time to an intrinsically useful product for his employer. How is the knowledge to be applied? The difficulty here is that in all the historiographic evidence from the development of science and technology there are almost no captive specimens of the transfer from science into technology. It is well known that such specimens can be bred by hindsight, and when this is needed to support science policy as in the famous reports, *Hindsight* and *TRACES*, it is obvious that the evidence and the conclusions are artifacts of the funding agency. More to the point, when one writes an honest and normal history of technology and one is at pains to explain why, when, and how changes in technology happened, the evidence is seldom if ever from causes within the development of science.

The conventional wisdom about the "obvious" fact that technology depends on the application of science has given rise to the misleading terms *applied science* and *applied research*, which lend themselves with great facility to the special pleading needed for funding arguments. Such ideas derive, I suppose, from the belief that technology such as radio communication must depend on a chain that runs back through Marconi and Hertz and leads ultimately to Maxwell's electromagnetic wave theory. Synthetic dyestuffs and modern pharmaceuticals can be traced back to the beginnings of organic and inorganic chemistry in the early nineteenth century. The trouble is that it seems very difficult to put one's finger on any incident where new scientific knowledge leads to an invention and then to its practical innovation.

Several lines of research into the nature of the innovative process show that most industrial innovations proceed more from the pull of the right market conditions than from the push of technological availability of a product. Indeed, much innovation comes from rather old technologies that have been waiting for somebody to spot them and to use them in manufacture. This has given rise to the term *almost technology* for describing something that we know, roughly speaking, how to do but have not yet put into practice. Even when some technique is known in principle, there can be a long and expensive process to get everything to come out in the desired fashion. The manufacture of the first atomic bombs is almost a classical story in this genre. At the very outset, the bomb was available as an almost technology because one knew what to do, more or less, to make it possible, in a way that a cancer cure is not at present, an almost technology. For the atomic bomb, a great deal of scientific and technological input was required to get the right materials into the right configuration in the right fashion, and the process of debugging was enormously costly and pressured even further by the urgency of the time factor. In another well-known historical example, how did it happen that thermodynamics owed much more to the steam engine than ever the steam engine owed to thermodynamics? In general, why is it that so much of so-called applied science is not the application of science to anything, but the scientific examination of artifacts and technical processes rather than at the natural world? Astronomy, physics, and chemistry are natural sciences, while metallurgy, electronics, and ceramic chemistry are applied sciences. It is as a consequence of this definitional status that in all the world's national statistics we find about two-thirds of the investment in applied science to one-third in basic science. The plain fact is that there are two parts of the artifactual subject matter of investigation for every one part of the natural.

Why does not science figure in the history of technology? How is it that so-called applied science is just as much a free good of international technical literature as basic science? The answer, I suggest, is that so often the history of science is treated as intellectual history, and so many historians of technology have been preoccupied by the necessary first stage of antiquarianism in which the artifacts have to be reduced to documents before the history can begin to be written. Most of our historians of science and virtually all the philosophers of science are former theoreticians rather than experimentalists.

It is clear that the junction of science and technology is in the interesting effects, strange substances, and new instruments of the experimentalist. Too often the philosophers of science have regarded scientific experiment as a means whereby theories could be confirmed, or with Popper, falsified, thereby reducing the scientific enterprise to a predominantly intellectual succession of theories. The true paradigmatic role is in the first instrument of all, the telescope of Galileo. It came into being because of the relatively recent availability of strong convex lenses that had just been produced by new lens-grinding machinery for the spectacle lens trade. The telescope gave Galileo data about the universe that nobody ever had before. What changed, in this first case, spectacularly, was the explicandum of science. It was not a question of mortal combat between the Ptolemaic and Copernican theories—the world of planetary phenomena was quite new. The artificial revelation of the telescope was such that all the people hearing about it needed to repeat the experience for themselves. Thus an optical instrument industry came into being and before very long gave birth to the microscope and thence to all the revelations of which that new wonder was capable. That, in turn, led in this crazily-convoluted chain of events to the discovery of biological cells and thence to the cure of dread diseases.

The telescope and microscope were the instruments that led to the change we call the Scientific Revolution. The role of the devices was not the testing of theories, but the revelation of new data that changed the world. The thermometer and the barometer provided measurement where there had been only qualitative data previously. The electrostatic generator machine and the vacuum pump produced conditions in the laboratory that nature had never provided for the observer. The chain of causality begins with a new technique, usually an elaboration of an old one, as with the lens craft of the optical instruments. This produces new data for science and some analysis of the craft itself. Thus the change in laboratory technique often leads both to an advance in science and to an improved technical availability.

From the beginning of the nineteenth century when the newly discovered force of voltaic electricity provided so many chemical reactions and physical phenomena, there developed a wealth of high technologies resulting from the newly available substances and effects. This move from low to high technology, from simple manufacture to complex production involving a high investment in people with advanced scientific and technical training, increased sharply during the world wars. After World War II with atomic energy and radar techniques, the quest for rocketry and space science, the sophistication of transistor chips and computers, this component of economy became dominant in the most developed nations and now increases the emphasis on the workings of the R&D investment which this chapter addresses.

With the social license for basic research, and the roots of scientific and technical invention and innovation now sketched in, we can proceed to a summary of a complete conceptual basis for matching the inputs to the outputs.

1. Most basic research and some applied research serves the main social purpose of preventing obsolescence of the personnel involved and provides them with access to the world's new knowledge in their domain of competence. The input amounts to about 30 percent of the resources used to provide a different nonresearch valuable product (such as teaching) from the personnel. The input may also include a certain cost of the experimental facility that must be made available for the research. The output is a free good of open publication in scientific and technical journals, reports, or books. This output may be measured bibliometrically in various ways and is roughly proportional to the input. Another by-product output may be a repertoire of experimental techniques and material that can eventually be drawn on to develop technologically in response to market availabilities.

2. Some predominantly applied research may arise in response to an essentially contractual relationship to supply valuable services, particularly in the professional service industries, public and private. In this case the research results in a product that is not published openly, but has particular benefit to those sponsoring the research. The input is given by the resources in personnel, funding, and facilities, and the output is the valuable service (such as airport siting advice, agricultural extension work) that was the subject of the contracting. Some of this applied research takes place in the area of investigation of new laboratory techniques that may be pushed toward recognized market possibili-

ties. The output from such effort is either valuable advice or a stock of almost technology.

3. Development arises in connection with the push from an almost technology into an established new product. In general, this involves much more input from production workers than from those with research front scientific and technical training. The input is the labor necessary to establish the new product in principle rather than in detail. The output is a new product available for market needs. There is a strong relationship between development input and the rate of product innovation or, as in some cases, the rate of increased productivity. The development investment as a fraction of the economic size of the industry increases as a power (perhaps the third power) of the growth rate in market output. In many industries, such as aerospace, which is the most development intensive, the rate of product innovation necessary for successful sales causes the R&D investment rather than vice versa.

Note

1. The background to the subject matter of this paper and related references to the scholarly literature are contained in what is in effect a preamble to this paper, "Toward a Model for Science Indicators," in *Toward a Metric of Science: The Advent of Science Indicators,* ed. Y. Elkana et al. (New York: John Wiley and Sons, 1978), chap. 4, pp. 69-95. A more explicit reworking of the concept that forms a bridge between that paper and the present one is "An Extrinsic Value Theory for Basic and 'Applied' Research," in *Science and Technology Policy,* ed. Joseph Haberer (Lexington, Mass.: D.C. Heath and Co., 1977), chap. 2, pp. 25-32. Finally, a fourth paper in this series, coming logically after the present paper, and applying its ideas to the inputs and outputs in the R&D economy of the United States, has been given in "Evidence: Joint Congressional Hearing, House Committee on Science and Technology, Oversight of National Science and Technology Policy, Organization and Priorities Act of 1976," Serial No. 95-77, Feb. 14, 1978, U.S. Government Printing Office, 73-79, and *Science and Public Policy* 5 (1978):190-195.

Index

Aarhus County, Denmark, 199, 206
Abdulkarim, A.J., 147
Abernathy, W.J., 100
Abramovitz, M., 171
Acquisitions, company, 59
Adams, J.G.U., 167-168, 221
Adaptive responses, 120-121
Administrative functions and administrators, 107, 113, 227, 254
Advertising, factor of, 40
Aerospace technology and aeronautics, 53, 205, 233, 260
Age: distribution factor of, 11, 20, 22, 209-210; process of, 69; of stock, 188
Agribusiness, 39-40
Agriculture, 39, 128, 181, 190, 215
Aircraft industry, 133-134, 172, 180-193, 196; technology, 58, 188
Allen, J.M., 8-9
Alternatives, evaluation of, 5
Ansoff, H.I., 8-9, 220-221
Anthony, R.N., 220
Antibiotics, 64, 84
Applied research, 215-216, 234, 259
Applied science, 200, 242-246, 257-258
Armour, H.O., 112
Army Research Office, 91
Arrow, K., 6, 112, 195
Ashly, Eric, 240-241
Assistance programs, 30
Astronomy, 132
Atomic energy, 259
Attention span, factor of, 5
Australia, 255
Automation, degrees of, 58
Automotive industry: foreign cars, 61; passenger cars, 57, 67, 82
Autonomy, full, 113, 116
Ayres, C.E., 132

Backward integration, 55, 59-60, 82
Baguley, R.W., 6
Bain, J.S., 145
Baldwin, W.L., 8
Bandwagon effects, occurrence of, 160
Bank interest, short-term, 27-29
Barn equipment, 41
Barth, Richard T., 2, 91-101
Basic: research, 234; science, 244, 256; skills, 254
Bay area south of San Francisco, 212
Bearing plants: ball, 57, 59, 67, 86; roller, 77

Behavioral theories and characteristics, 5, 167
Bell, Daniel, 246
Bertalanffy, 138
Bessemer Converter, use of, 136-138
Bioanalysis, factor of, 72
Biochemistry, 40
Biology, 63, 85, 176, 241; and evolution, 131; research in, 255-256; science of, 146
Bonds, government interest on, 14-15, 26-29
Booth, J.J., 6
Boston Route 128 firms, 205, 210-211
Bottlenecks and bottoms-up technology, 172, 200
"Bounded rationality" defined, 110
Bright, J.R., 6-7, 9
Bronowski, Jacob, 240-241
Brooks, H., 6-9, 217-219
Bruun, Michael O., 199, 203-212
Budgets and budgetory allocations, 2, 54, 68-69, 76, 98
Bupp, I.C., 149
Bureaucracy, procedures of, 43-44, 78
Business: cycles, 59, 65, 68, declines in, 67; studies, 40
Butler, Samuel, 138

Cameron, T.A., 5
Campbell, D.T., 99
Canada, 12, 26-27, 255; Ministry of State for Science and Technology, 10
Cantley, Mark F., 127, 143-150
Capital: availability of, 206; costs, 145; equipment, 15, 155; goods, 76, 106; investments, 76, 217; producing of, 128-129; stock, 188
Career orientation, 207
Cartels, 105
Cash flow, problems of, 145
Cause-and-effect theory, 94
Cavendish, Henry, 252
Centralization policies, 63, 113
Channon, Derek F., 73
Chemical Corporation (Ch-Corp), 53, 57-58, 60, 62-64, 66, 68, 78, 81, 83
Chemical industry, 39, 58, 60, 85, 146, 149, 203
Chemistry and chemists, 57, 63, 71
Chicago school of thought, so-called, 105
Childs, G.L., 8
China, 251

Chomsky, Noam, 240
Civil service, 253
Clarity, element of, 94, 98
Coding systems, 91, 112, 120
Cold War, effects of, 240
Comanor, W.S., 8
Commercial markets, 203-205, 211
Communication(s) network, 3, 91, 112-113 175, 257; intergroup, 92-95, 99; organizational systems, 119; variables of, 92-93, 120
Competition, degrees of, 69, 168
Competitive: behavior, 232, 236; conditions, 144-145; market, 200; pressures, 160-164
Competitors and competitiveness 68, 77; foreign, 39, 72; national, 248
Computers and computer industry, 85, 128, 180, 182, 191-193, 204, 259; digital, 171-172
Conformity dimensions, 92, 98
Confrontation (CONFR), and problem solving, 94-98
Conservation policies, 138
Constraints: environmental, 144; financial, 99; supply, 138
Consultants, 2, 50, 72
Consumers and consumer markets, 60, 118
Contract: companies, 206; government, 15, 26-29; industrial, 205
Congrol groups, use of, 206
Cooper, A.C., 8-9, 212
Cooperation, need for in success, 44-50
Copernican theory, 258
Corporate: behavior, 53; development, 2, 69-73; diversification, 59, 77-78; growth, 55-62; laboratories, 87; managing directors, 53; organizations, 107; profitability, 67
Corporate Strategy (Ansoff), 221
Costs: capital, 145; labor, 65, 133; transportation, 72
Councils for Scientific and Industrial Research, 255
Coupling agent, 99-100
Cranston, R.W., 9
Creativity, human, 251-252
Crops: rotation system for, 175; types of 177
Culture, factor of, 248, 251-252, 255
Culture Against Man (Henry), 240
Customer: demand, 42; problems, 76; processes, 85; technologies, 68
Cybert, R.M., 5, 8

Dairy industry, 77, 85
Davies, Stephen, 127, 153-169
De Bono, Edward, 240, 242, 244
Decentralization, policy of, 63, 113, 144
Decision making: individual, 1; intergroup, 91-95; variables relevant to, 5-6, 127, 224
Defense programs, 56, 215, 218, 227; research on, 219-220, 226
Demand: customer, 43; -oriented theory, 76, 138
Demand-pull technology hypothesis, 2, 116-118, 144, 190
Demography, factor of, 177, 226
Denison, Edward F., 217
Denmark, 199, 206
Dependence, policies on, 68, 235-236
Depreciation, physical, 15
Derian, C., 149
Deutsche Shell Chemie, 149
Developing countries, 194, 259
Development: advanced, 114; corporate, 2, 69-73
Diffusion: element of, 199; information, 165; of know-how, 139-140; and speed concentration, 156-165, 169; time path of, 163
Digital computers, 171-172
Dirlam, J., 167-168
Discrimination, group, 18-20
Disease: areas of, 67, 252; contagious, 160; cure of, 258
Diseconomies of scale, 147
Disman, S., 9
Displaced time series, 75
Distribution, 56, 145; firm size, 165; population, 10
Diversification concept: conglomerate, 55; corporate, 59, 77-78; defined, 54-55; degree of, 2; patterns of, 60, 70; synergistic, 81
Dobrov, G.M., 233-234
Douds, C.F., 93
Drucker, Peter F., 7, 245-246
Dullemeijer, P., 176
Durbin-Watson (D-W) test statistics, 159, 184, 187, 189
Durrer's experiments, 136-138
Dynamic efficiency, 116, 119

Economic: conditions, 14, 18, 26; development, 216-217; environment, 30; geography, 226; growth, 171, 225; incentives, 5; investments, 221; trends, 10
Economies of scale, 127-128, 134, 144-149
Economy, 131-132, 176; classification in the, 138-139; industrial, 153; market, 144; national, 143, 234; socialistic, 144
Education: factor of, 20, 22, 25, 212;

Index

formal, 11; levels of, 140, 207, 217, 231; postgraduate, 11
Edvardson, B., 1, 39-50
Electricity and electrical appliances, 63, 150, 172
Electronic: companies, 210; control, 137; industry, 36, 63, 67, 69, 203, 206, 258; systems, 58, 206
Electronics Corporation (El-Corp), 53, 57-58, 61, 66, 70-71, 81, 83
Electronics Corporation subsidiaries (El-Corp Subs), 53
Employees: loss of, 72; number of, 44, 48-50, 57
Employment and programs for, 21, 24, 27, 29, 216
Energy, 138; atomic, 259; hydro-electric, 81; prices for, 225; requirements, 15, 26
Engineering and Food Corporations (E&F-Corp), 53, 55, 57-58, 60-64, 66, 68, 72, 74-78, 81, 83, 85
Engineering and Steel Corporations (E&S-Corp), 53, 56-58, 60-66, 68, 70-73, 76-78, 81-83, 86-87
Engineering systems, 3, 57, 61, 63, 69, 114, 140, 172
Engineers, profession of, 92, 134, 149, 245, 253-254
English-language taxonomy, 239, 242-243, 246-247
Entrepreneurship and technical changes, 206-212, system of, 256
Environment, 132, 221; constraints, 144; economic, 30; physical, 232; task sources, 5, 91
Equipment, 173; barn, 41; capital, 15, 155; physical, 143; transportation, 171, 196; utilization of, 137
Essays in the Theory of Risk Bearing (Arrow), 112
Ethylene plants, 149-150
Europe, continental, 139
Evaluation projects, 14, 16-18; standards of, 26, 71
Evan, W.M., 99
Evolution: self-perpetuating process, 132; technological, 138-139
Exchange: of co-workers, 205; foreign rates of, 26-28
Executives, attributes of, 10-11, 25-30
Experience, factor of, 131
Exploratory research, 3, 114
Eyring, H.B., 100

Farm: machinery industry, 1, 3, 39-43, 50, 171, 190, 196; sizes, 184; tractors, 174, 180-181, 185, 188-193

Farming and farmers, occupation of, 85
Feasibility studies, 17
Feedback, 2, 100, 187
Fellner, W., 6
Fertilizers, use of, 57-58
Feyerabend, P., 133
Financing, 55-56; constraints, 99; short-term, 25
Firms: characteristics of, 10-15; size of, 128, 153, 165
Fisher, J.C., 147, 149
Food companies, 40
Forcing (FORCI), 94-97
Foreign: car industry, 61; competition, 39, 72; control, 12; exchange rates, 26-28; market, 54; subsidiaries, 62; technology, 61, 72, 81; trade, 139
Fores, Michael, 200, 239-249
Forests and forestry products, 41, 58, 60, 67, 70
Forward integration, 55, 59-60
Foster, R.N., 6-7, 9
Friedman, G., 150
Funds and funding, 99; agency for, 257; direct, 30; government, 17-18; of research, 215; reform of, 256

Galbraith, John K., 6-8, 15, 26, 43
Garden machines, 41
General Electric Company, 147, 149
General Motors Corporation, 147, 149
Geography: economic, 226; factor of, 144, 177, 224
Geological Survey of Sweden, 216
Germany, 86; language of, 243, 246
Gerstenfeld, A., 8-9
Glagolev, Vladimir N., 127, 143-150
Goal(s): achievement of, 115, 119; group, 116; performance, 99; short-term, 115-116; societal, 222
Gold, B., 145-147
Gompertz function, 178
Gomulka, St., 139
Goodman, L., 101
Goodman, R.A., 100
Goods, capital, 76, 106
Government: bonds, 14-15, 26-29; contracts, 15, 26-29; expenditures, 14-15; funding, 17-18; laboratories, 212; subsidies, 15, 18; support role for research and development, 14-18, 25-26
Grandstrand, Ove, 2, 53-89
Grass-roots research, 62-78
Great Britain. *See* United Kingdom
Green revolution, effects of, 175
Griliches, Z., 156

Gross national product (GNP), 15, 27, 203, 224-225
Group(s): control, 206; effectiveness of, 93; goals of, 116; managers of, 98; performance of, 92, 97, 101
Growth concept, definition of, 54

Hackman, J.R., 100
Hamberg, D., 6-9
Hansen, F.H., 149
Harvard University, 105
Hayami, Y., 6
Health, factor of, 215
Henry, Jules, 240, 242, 248
Hicks, John 196
Hierarchical sectors, 119
Hilberg, David, 132
Hindsight report, 257
Hirschmann, A.O., 138
Holmström, S., 39
Horizontal integration, 55, 59-61
Hospitals, 67
Housing sector, 215
Huettner, D.A., 146
Human: creativity, 251-252; information-processing capability, 1; techniques, 131
Hutton, S.P., 245
Hydroelectric energy sources, 81
Hypertension, effects of, 88

Immunology, 40
Implementation, range of, 111-114, 118, 120, 200, 232; threshold issue, 112-113, 233
Incentive systems, 113, 116; economic, 5; tax, 26; technological change, 109-110
Independent companies and policies, 206, 209-211, 235-236
India, 255
Industrial: complexes, 143; concentration, 156-160; contracts, 205; countries, 194, 203; economies, 153; input-output system, 55; organizations, 105-106; realities, 145; revolution, 138; sectors, 200; techniques, 132
Industries: high-technology, 121; inter-processing differences, 159; service, 181
Industrispeak: dialect in English, 245-247; tradition, 239, 242
Industry, Trade, and Commerce, Department, Canada, 5
Inflation, 15, 57, 74
Information, 6, 10, 14-18; diffusion, 165, 205; exchange of, 117; flow of, 115, 157, 164; preferences, 5, 15-18, 25
"Information basked" methodology, 5-11, 13

"Information impactness," definition of, 110-111, 117
Innovation, 165-168; activity, 2, 171; process of, 137, 180-181, 199; radical, 3, 127; technological, 3, 40-43, 48, 53, 145; in transportation, 190
Innovativeness (INNOVA), 8, 17, 94-98
Institutes, research, 50
Integration: backward, 55, 59-60, 81-82; forward, 55, 59-60; horizontal, 55, 59-61; organizational, 91; vertical, 65, 70, 105, 111, 113-115, 119-120
Interdependencies, task, 91-92, 99
Intergroup: climate efforts, 3, 92, 100; communication, 92-95, 99; decision-making, 91
Internal Journal of Systems Science, 183, 186
International Business Machines, computers by, 174
International Institute of Applied Systems, 149
Internationalization, factor of, 54, 67-68, 72, 78, 81, 87-88
Interorganizational systems, 3
Interpersonal problems, 99
Interview: data, 92-93; notes, 98; personal, 42; telephone, 41, 43
Intraorganizational systems, 3
Inventions and inventors, 50, 60, 72
Inventory trends, 25-29
Investigation, empirical, 5
Investment(s): capital, 76, 217; character of, 193; climates, 25; decisions, 17; economic, 221; gross, 128, 188; opportunities, 30, rate of, 137; respect to, 231; road, 225
Irrationality, factor of, 132
Irrigation systems, 175

Japan, 136-137, 139
Jennergren, C.G., 227
Jewkes, J., 43

Kaldor, N., 134
Kalecki, M., 138
Kamien, M.I., 6, 166
Kennedy, C., 166
Kepler's Harmony of the world thesis, 133
Keynes, J.M., 6-7
Kimberley, J.R., 176
Klein, Burton, 115-116, 119
Know-how, diffusion, 60, 139-140, 173, 203-204
Kondratieff cycle, 203
Kotler, P., 8-9
Kuhn, Thomas S., 133

Index

Kuznets, Simon, 240, 243-244

Labor: availabilty of, 144; cost of, 65, 133; dependence on, 68; force, 173, 193; hours of, 221, 225; input, 134, 172; manual, 39; productivity, 91, 137, 190; shortages of, 39
Laboratories, 62, 70; corporate, 87; directors, 92; governmental, 212; Massachusetts Institute of Technology (MIT), 206; materials, 83; research 83, 252; sizes, 71; source, 211; techniques, 259
Landsberg, A., 5
Langrish, J., 245
Language: English-speaking, 239, 242-243, 246-247; German, 243, 246; use of, 119-120
Lawler, E.E., 100
Lawrence, P.R., 93, 98, 245
Layton, Edwin, 240
Leadership style, 99-100
Learning: effects of, 128, 137; function of, 134, 190-195; individual, 72; process, 131, 134; role of, 129, 184, 194; scope of, 128; short and long term, 127
Learning-by-doing hypothesis, 128, 133, 138, 173, 181-184, 187, 190, 193, 195
Learning-via-diffusion hypothesis, 187-188
Lee, T., 149-150
Leibnizian academies, philosophy of, 255
Leonard, W.N., 6, 8-9
Licenses, need for, 68
Life cycles and styles, 18, 26, 58, 64, 254
Light-water nuclear reactors, 149
Likert scale, 12, 92-93
Lindblom, C.E., 138
Linde process, 136-138
Lindström, C., 43-44
Linz, 137
Liquidity, element of, 8
Lithwick, N.H., 8
Location: factor of, 20, 23; geographical, 144; and sales, 29
Locomotion, techniques of, 131
Locomotives, stock of, 180, 185-193
Lorsch, J.W., 93, 98
Loyalty, individual and organizational, 91, 113
Lucas, N.J.D., 147

Machine-building sector, 187, 193, 195
Machines: design of, 39, 172; paper, 155; and tool industry, 128, 190
Macroeconomics, 54
Maddison, A., 139
Maier, H., 233-234
Malinowski, Bronislaw, 241-242

Management: commitments, 62; levels of, 70-71, 143, 172; senior, 25; top, 53, 60, 77
Managers and managerial actions, 53, 70, 91-92, 99, 113; group, 98; utility plant, 149
Manpower, 98, 217
Manufacturing, 3, 172
Mansfield, Edwin, 8-9, 153-156, 159-165, 204, 246, 249
Mant, Alistair, 245
Manual labor, 39
March, J.G., 5, 8
Markets and marketing, 10, 23, 27, 53, 55, 145; activities, 3; changes in, 58-59; characteristics in, 15; commercial, 203-205, 211; competitive, 200; consumer, 60, 118; economies, 144; external, 105; failures of, 116, 120; foreign, 54; metals, 59; mechanism, 107; oligopolistic, 145, 167; organizations, 210; performance, 105-106; and production, 62, 114; stability, 15; technical, 42; variables of, 40; volatile, 13-15
Marquis, Donald G., 204
Marshall, George C., 139
Mason, Edward S., 105
Massachusetts Institute of Technology (MIT), 206
Mathematics, 132; Maxwell's electromagnetic wave theory, 257
McGlauchlin, L.D., 9
Mechanics and mechanization, 39
Medicine and medical care, 67, 71, 77
Mertonian line of sociological enquiry, 251
Metallographic Institute, 216
Metallurgy industry, 258
Metaphysical concepts, 133
Metatalk in accounts, 245-247
Meyer, 100
Microcomputers, 63
Microeconomics, 54
Milking machines, 41, 57
Mines and mining, 60
Mining and Metallurgy Corporation (M&M Corp), 53, 56-58, 60-64, 66-68, 78, 81, 83
Mission-oriented basic research, 216, 218, 226; projects, 98
Mitchell, V.F., 5
Mobility, advantage of, 72
Monopoly and monopolists, 168
Morale, factor of, 72
Morgan, P., 10
Morrison, D.G., 101
Morton, J., 100
Motivation, entrepreneurial, 207

Mottley, C.M., 8-9
Multinational corporations, 2, 53
Mumford, Lewis, 241-242, 244
Myer, H.H., 204

National: culture, 248; economy, 143, 234; railroads, 220; science policy, 215
National Aeronautics and Space Administration (NASA), 91, 205
National Defense Research Institute, 216
National Institute for Plant Protection, 216
National Road Research Institute, 216
National Science Foundation, 91, 254
National Veterinary Institute, 216
National sciences and scientists, 203, 217, 239-240, 243, 249, 258
Navigation and navigational facilities, 132, 177
Needham, Joseph, 251
Nelson, R.R., 9, 120, 171
Newton, R.D., 8-9, 132
Nilsson, B., 39
Nobel, Alfred, 86
Nonprofit organizations, 212
Northwestern University, 91, 101
Nutrition and nutritional products, 89
Nyström, Harry, 1, 39-50

Occupational abilities and specialities, 5, 175
Ohm, George Friedrich, 252
Oligopolistic: industries, 167; organizations, 105; producer markets, 60, 145, 167
Olley, Robert E., 105
Opportunities and opportunism, 114-117; behavior, 113; defined, 110-111; investment, 30; and profits, 67
Ore deposits, 61
Organization for Economic Cooperation and Development (OECD), 216-220
Organization for Economic Science policy field, 216
Organizations and organizational factors, 105-106, 109-120; attributes of, 10; integration, 91; loyalties, 113; marketing, 210; nonprofit, 212; oligopolistic, 105; permeability, 3
Output per capita, factor of, 171
Overoptimism, problem of, 60
Ownership, patterns, of, 145
Oxygen steel process, 167

Paper: industry, 70, 81; machines, 155
Pareto distribution, 138, 195
Patents: acquisition of, 60, 68; pools, 117; products, 50; protection of, 42-43
PATH models and coefficients, 94-97

Payback period, 1, 9, 18
Pelc, Karol I., 111-112, 200, 231-236
Perceived communications problems (PCP), 92-99
Performance: evaluation criteria, 26; failure, 106, 120; goals, 99; group, 92, 97, 101; market, 106; mechanisms, 106, 145; operational, 94; reward expectancies, 100; task, 91; team, 99; threats to, 120; unit of efforts in, 93
Personnel in research, 60, 119
Pessemier, Edgar A., 204
Peterson, R.W., 9
Petroleum, refining of, 172
Pharmaceutical Corporation (Ph-Corp), 53, 57-58, 60-68, 71, 74-77, 82-83, 87-89; subsidiaries, 53, 72, 77-78, 88
Pharmaceuticals and pharmaceutical industry, 39, 58-59, 63, 76, 165; research in, 71-72, 87-88
Pherson, M., 235
Phillips, Almarin, 3, 9, 105-129
Physical: environment, 232; equipment, 143; sciences, 146
Physicians, profession of, 67
Physics, solid-state, 58, 71
Planck, Max, 253
Planning: long-range, 226-228; policies, 199-200; strategic, 220-226; technological, 199-200
Plant layouts, 172
Pluralism, practice of, 217
Poisson distribution, 252
Policy makers and making, 54, 77, 227
Politics and politicians, 205, 255
Pollution: control of, 15, 17, 27; reduction in, 224-225
Polymer technology, 64
Popper, 258
Population: distribution, 10; growth, 27, 29, 221
Postgraduate training, 11
Pressures: competitive, 160-164; political, 205; on prices, 69
Preston, R.S., 6
Price, Derek de Solla, 200-201, 243, 248, 251-262
Prices: effects of, 40; falling, 144; pressures on, 69
Producer market, 60, 145, 167
Product(s); changes, 56-58; life cycles, 70; line levels, 54-55; and marketing development, 114; new, 47, 50; nutritional, 89; patented, 50; sales structure, 57; substitutions, 57
Production: decentralized, 144; facilities, 62; and marketing, 62, 114; problems in,

Index

55-56; processes and methods, 111, 116, 119, 135, 173; responsibility, 219; services, 42, 222; skills, 201; techniques, 132; technology, 58, 77
Productivity: changes, 15; growth of, 127, 133-134, 171; labor, 91, 137; share of, 134
Professional: organizations, 116-118; skills, 254
Profitability, 65; changes in, 73; corporate, 67; and sales, 14-18
Profits: average, 14, 28; maximization, 113; opportunities, 67, 69; rate of, 15, 18, 138
Program responsibility for research, 218-219
Project selection, 5
Ptolemaic theory, 258
Public: research institutions, 211-212; services, 223
Pulp and Paper Corporation (P&P Corp), 53, 57, 60, 62-64, 66-68, 76, 78, 81, 83
Pure science, 244-246
Push-pull dichotomy pattern, 2, 61, 79, 116-119, 144, 190

Qualified technicians, 173
Questionnaires, use of, 10, 92
Quinn, J.B., 6, 8-9

Radicalism and innovations, 3, 127
Railroad systems, 187, 204; national, 220
Randolph, R.H., 234
Rasmussen, P.N., 203
Rates of return, influence of, 1, 162
Rationality, procedural, 1, 144, 160
Rauch, W.D., 234
Raw materials: factor of, 60-62, 68, 81; sources of, 72; supplies of, 231
Recreational items, 58
Recruitment, element of, 72
Reduction of target date uncertainty (RETARU), 94, 96-97
Reduction of technical uncertainty (RETECU), 94, 96-97
Regional Dissimination Center program, 205
Regional imbalances, factor of, 224
Religion, concepts, of, 133
Research: basic, 234; biomedical, 255-256; exploratory, 3, 114; funding of, 215; institutions, 2, 50, 211-212, 215; laboratories, 83, 252; sectoral, 215-220
Research for the Defense Sector report, 218
Research and Development (R&D), 53-55, 58-59, 61-64, 67-78, 83, 85-88
Resources, allocation of, 68, 218
Responsibility, trait of, 92, 98, 218-219
Revolution, industrial, 138; scientific, 258

Reward system, 117-118; performance expectancy, 100
Rey, Lars, 245-246
Reynolds, L.G., 196
Risk(s), taking of, 15, 44, 94, 98, 100   m
Roads and waterways, system of, 175, 220, 225
Roase, Ronald H., 105
Roberts, Edward B., 205-207, 210-211
Romeo, A., 159-161, 164
Rosenberg, N., 6
Rothchild reform of funding, 256
Ruttan, V.W., 6

Sahal, Devendra, 128, 146, 171-197, 235, 243, 245
Sales and selling structure, growth of, 3, 14, 18, 20, 23, 26, 28-29, 41-42, 57, 74
Salter, W.E.G., 196
Salveson, M.E., 78
Sawyers, D., 43
Scale: changes in, 129; economies of, 127-128, 134, 144-149; levels of, 143-145
Schenk, W., 137
Scherer, F.M., 8-9, 43, 145
Schmookler, J., 6, 79, 196
Schon, Donald A., 204
Schumacher, E.F., 138
Schumpeter, Joseph, 6, 8
Schwartz, Sandra L., 5-30, 166
Schwarz, B., 200, 215-228
Science: advancement of, 68-69, 92; applied, 200, 242-244, 257-258; basic, 244, 256; biological, 146; development of, 132, 135; interaction of, 131; logic of, 132; national policy, 215; natural, 203, 217, 239-240, 243, 249, 258; physical, 146; pure, 244-246; research, in 252; role of, 251; social, 217, 243; and technology linkage, 10, 53, 242, 246; trouble with, 242-245
Science, Growth, and Society (Brooks' Committee), 217
Science Research Council, 243
Science and Technology in the Management of Complex Problems, report, 219
Scientific: developments, 119, 138; disciplines, 115-116; journals, 252-253, 256, 259-260: revolution, 258
Scott, Bruce R., 72-73
Self-adjustment, 86
Self-employment, 207, 210
Self-perpetuating process, 132
Self-reinforcing factors, 69-70, 76
Service: industries, 181; professions, 253; and production, 42, 222; public, 223

Shani, M., 220
Shipbuilding industry, 61, 172
Simmonds, W.H.C., 146-147
Simon, Herbert, 1, 5, 105
Size: company, 43-46, 50, 176; difficult to assess, 70; laboratory, 71
Skill(s): acquired, 241; basic, 254; production, 129, 201; professional, 254; technical, 208, 244; types of, 131, 194, 246
Smith, V.K., 7, 134
Smoothing (SMOOT), 94-97
Snow, C.P., 240, 242-243
Social: background, 206; marketing strategies, 30; order, 240; organization, 232, 252; sciences, 217, 243; values, 255
Socialist: countries, 255; planned economies, 144
Society, 131-132, 255; changes in, 40; goals, 222-223; implications for, 144; needs of, 217; permeable, 140; pressures, 68, 240; technology in, 251; traditional 232
Sociology, issues of, 150, 176, 251
Solo, R., 196
Solow, 171
Sorge, Arndt, 247
Source materials, 72, 144, 211
Soviet Union, 255
Space program, 205, 212
Specialization and specialities: factor of, 68, 144; occupational, 175; and scale hypothesis, 128, 176, 184-187, 190, 193
Speed, concentration and diffusion, 156-165, 169
Spinoff: companies, 205-206, 210-211; processes, 78, 212
Stagnation, element of, 178
Standard of living, 216
Standard operating procedure (SOP), 1, 5, 25-26
Statistics: Durbin-Watson, 159, 184, 187, 189; traits of, 12
Steel: converters, 137; industry, 39, 57, 67, 77; production, 86
Steindl, Josef, 127, 131-141
Stewart, J.M., 8-9
Stocks: capital, 188; and market trends, 26
Strategic planning for transportation, 220-226
Strategy and Structure of British Enterprise, The (Channon), 73
Subsidiaries, 53, 82; decentralized, 87-88; of Electronics Corporation, 53; foreign, 62; Pharmaceutical Corporation, 53, 72, 77-78, 88
Subsidies: forms of, 128, 195; governmental, 15, 18
Substitutions; 67, 71, 74; product, 57; technology, 56, 78

Supervisors, technical, 93
Supply: constraints, 138; -oriented theory, 76; raw material, 231; shortages of, 56
Surveys and surveyors, 15, 29, 132, 216
Sussex science policy group, 231
Svensson, J.E., 215
Sweden, 39, 41, 78, 215-216, 220; multinational corporations in, 2, 53; security policies of, 218, 227; transport system in, 224
Swedish Agricultural Board, 41
Swedish Meteorological and Hydrological Institute, 216
Synergy and synergistic diversification, 40, 43, 60, 64-67, 81, 89
Systems Assessments of New Technology (Dobrov et al.), 234

Taking of risks, 15, 44, 94, 98, 100
Tank ships, 177, 180-182, 184, 188-193
Tariffs: favorable, 15; issue of, 128, 194
Task: interdependent technical groups, 91-92, 99; performances, 91
Taxes and taxation policies, 14-18, 26, 29
Taxonomy, question of, 145, 239-249
Team performance, unity efforts, 99
Technical: change, 25, 168, 206-212; expertise, 91; journals, 256, 259-260; marketing, 42; progress, 127, 131, 135; risk-taking, 100; skills, 208, 244; success, 1; supervisors, 93
Technicians, qualified, 173
Techniques: industrial, 132; new, 127-128, 131; of transportation, 131
Technoeconomic situations, 235
Technology: bottoms-up, 200; changing, 3, 17, 65, 106, 190; customer, 68; evolution of, 138-139; external, 59; foreign, 61, 72, 81; innovations, 3, 40-43, 48, 53, 145; interaction of, 135-138; mature, 69; packaging, 64; planning, 199-200; polymer, 64; production, 58, 77; push-pull hypothesis of, 2, 116-118, 144, 190; and science linkage, 10, 53, 242, 246; in society, 251; sophisticated, 162; strategy, 232-234; substitution of, 56, 78; synergistic use of, 40, 43; tank ship, 184; tractor, 184, 194; transfer process, 203-205; in transportation, 128, 176, 190-191; utilization program, 205
Technology and the Academics (Ashly), 240
Teece, David J., 105, 112
Teitel, S., 235
Telephone, use in interviewing, 41, 43
Thirlwall, A., 166
Thomas process, 137
Thurston, R.H., 7
Tilles, S., 8-9

Tilton, J.E., 6, 8
Time; element of, 135; lag in, 79; path of diffusion, 163
Tomlinson, J.W.C., 5
Topography, factor of, 177
TRACES report, 257
Trade: foreign, 139; workers, 131
Training; financial, 117; postgraduate, 11, vocational, 173
Transfer process, technology in, 203-205
Transport Research Delegation in Sweden (TRD), 215, 220, 225, 227
Transport Research Prospective plan (TRP), 220-221, 224, 226-227
Transportation Corporation (Tp-Corp), 53, 56-58, 61, 63-66, 69, 71-73, 77, 82-83; subsidiary of (Tp-Subs), 53, 70, 77
Transportation system, 192-193, 215; agricultural, 41; costs of, 72; equipment for, 171, 196; innovations in, 190; and road agencies, 220; and society, 223; strategic planning for, 220-227; techniques of, 131; technology, 128, 176, 190-191; urban, 256
Trail and error, process of, 131
Two-thirds power law, 145

Unemployment: levels of, 26-27, 231, 256
United Kingdom, 127, 147; manufacturing industries in, 153-154; science policies of, 255-256
Unity of effort (UNIEFF), 93-97

Universities, influence of, 2, 50, 60, 72, 212, 255
Urbanization: growth of, 144; structure of, 221; and transportation, 256
Utility plants, managers of, 149

Varhinger philosophy, 251
Varimax rotation system, 14, 19, 93
Verdoorn, P.J., 134
Vertical integration, 65, 70, 105, 111, 113-115, 119-120
Vertinsky, Ilan, 1, 5-30
Viggen combat aircraft, 70
Vocational training, 173

Wage settlements, 25-29
Warmth, interteam element of, 94, 98
Water, control of, 175
Weiss, L., 156, 166
Williamson, Oliver E., 8, 105, 106, 110-111, 113, 120, 165, 168
Winter, S.G., 120, 171
Wood products, 57
Workshop and workers, 128, 139, 246

Yale University, 101
Young, A., 134
Yule, G.U., 138

Zipf, G.K., 139
Zuckerman, Sir Solly, 240-242, 244

About the Contributors

Richard T. Barth received the Ph.D. in organization theory from Northwestern University. He joined the Faculty of Commerce and Business Administration of the University of British Columbia in 1970. He is on the editorial board of *IEEE Transactions on Engineering Management*. His current research interests include the organization and management of interdisciplinary R&D units, the implementation of OR-MS, and an assessment of research-on-research studies. His publications have appeared in such journals as *R&D Management, Socio-Economic Planning Sciences, Organizational Behaviour and Human Performance,* and *Human Relations*.

Michael O. Bruun received the Ph.D. in semiconductor physics and technology from Stanford University in 1969 and was a research associate at W.W. Hansen Laboratories of Physics there from 1969 to 1970. He joined Physics Lab III of the Technical University of Denmark in 1972 as assistant professor. He is presently associate professor at the Institute of Management, University of Aarhus, Denmark. His publications include *A Study of Economic Follow-Effects caused by the Industrial Contracts of a Large Basic Research Center* (1977) and *Danish Entrepreneurs* (1978).

Mark F. Cantley received the B.A. in mathematics from Cambridge University in 1963; a diploma in operational research in 1965; the M.Sc. in economics from the University of London in 1965; and a certified diploma in accounting and finance in 1978. His scientific interests are in corporate and strategic planning. At present he is working in the management and technology area of the International Institute for Applied Systems Analysis in Austria. He has published widely and is a member of the United Kingdom Operational Research Society for Long Range Planning.

Stephen W. Davies received the Ph.D. from Warwick University in 1976. A lecturer in economics at Sheffield University, he has also worked as research officer at the National Institute for Economic and Social Research in London (1970-1972). He has published articles on forecasting, industrial concentration, and technical progress, and is the author of a book on the diffusion of new processes.

Bo Edvardsson is research associate at the Institute for Economics and Statistics of the Swedish University of Agricultural Sciences in Uppsala. Together with Harry Nyström he has studied company strategies for research and development in Swedish Agribusiness firms.

Michael Fores worked for a number of years as an engineer and has more recently been a member of the British Government Economic Service, where he has worked on economic aspects of technical change. He is coeditor of *Manufacturing and Management* (1978).

Vladimir N. Glagolev received the first diploma in 1962 from the Kaunas Polytechnical Institute and the diploma of doctor of economic sciences in 1972 from Vilnius University, where he specialized in economics, organization of management, and planning of the national economy. In 1978 he joined the International Institute for Applied Systems Analysis, working in the management and technology area. He is working on practical problems of organization and management and is interested in the application of systems analysis in economics, management and planning, and organization design. His publications deal with management and planning of technological change, mechanization and automation in industry, the interbranch territorial program of industrialization of auxiliary production, and the effectiveness of science and technological progress in the national economy of the Lithuanian SSR.

S. Ove Granstrand received graduate degrees in mechanical engineering from Chalmers University of Technology; in mathematics from the University of Gothenburg; in economics from the Graduate School of Business in Gothenburg; and in operations research from Stanford University. He received the Ph.D. from Chalmers University of Technology in R&D management and technological innovation, which is his main field of research interest. His professional experience also includes teaching and consulting in government and industry.

Harry Nyström is professor of marketing and organization theory at the Institute for Economics and Statistics of the Swedish University of Agricultural Sciences, Uppsala. He was previously a research associate at the Stockholm School of Economics and professor of business administration at the University of Uppsala. His main research interest is in product development and organizational innovation. He has published several books and articles on the management aspects of innovation, most recently *Creativity and Innovation* (1979).

Karol I. Pelc received the Ph.D. in electronics from Uppsala University and the Ph.D. in business administration from the Technical University of Wróclaw. He is director of the Forecasting Research Center at the Technical University of Wróclaw and a member of the Poland 2000 Committee of the Polish Academy of Science. His professional activity is concerned with R&D management and technological forecasting.

Almarin Phillips is professor of economics and law at the University of Pennsylvania. He received the Ph.D. from Harvard University in 1953 and has been a visiting professor at the University of Hawaii, Ohio State University,

McGill University, Warwick University, and the London Graduate School of Business Studies. He was consultant to the U.S. secretary of the treasury from 1973 to 1977, a senior fellow at the Brookings Institution from 1970 to 1975, codirector of the U.S. President's Commission on Financial Structure and Regulation from 1970 to 1971, and a member of the National Commission on Electronic Funds Transfer from 1975 to 1977. He has published more than seventy papers and several books, including *Technology and Market Structure: A Study of the Aircraft Industry* (Lexington Books, 1971) and *Promoting Competition in Regulated Markets* (1975).

Derek deSolla Price is Avalon Professor of the History of Science at Yale University. He received a doctorate in experimental physics in London and a second doctorate in the history of science from Cambridge University. He is active in research on historical studies of ancient science, the history of scientific instruments, the archaeology of science, and the relations between the social and the natural sciences. In 1956 he was brought to the Smithsonian Institution to help plan the Museum of History and Technology. Following that he became a Fellow at the Institute for Advanced Study, Princeton. He has published some two hundred scientific papers and six books, including *Science Since Babylon* (1961 and 1975), *Little Science, Big Science* (1965), and *Gears from the Greeks: The Antikythera Mechanism, A Calendar Computer from ca. 80 B.C.* (1974-75). He is also co-editor of *Science, Technology and Society: A Cross-Disciplinary Perspective* (1977).

Sandra L. Schwartz is an assistant professor of policy analysis at the University of British Columbia. She received the M.A. in economics from the University of Wisconsin and the Ph.D. in policy analysis, decision sciences, and resource management from the University of British Columbia. She has participated in research projects at the University of California, Berkeley, and at Stanford University. Recently she directed a large international project on R&D investment policies supported by NATO and the Canadian Department of Industry, Trade, and Commerce. She has published articles in the areas of economic development, planning, and multicriteria decision making.

Brita Schwarz received the Ph.D. in mathematics from Stockholm University in 1958. In 1964 she became head of the Division for Systems Analysis at the Research Institute of National Defence; and in 1967, she was consultant to the Organization for Economic Cooperation and Development in Paris, specializing in program budgeting and long-range planning. She is currently director of research at the Economic Research Institute of the Stockholm School of Economics. She is the author of numerous reports and publications in systems analysis and long-range planning and editor of *Budgeting, Program Analysis and Cost-Effectiveness in Educational Planning* (1968) and *Operations and Systems Analysis* (1969).

Josef Steindl received the Ph.D. from the University of Vienna. He began his career at the Austrian Institute für Konjunkturforschung. A research lecturership at Balliol College, Oxford, took him to England, where he subsequently worked at the Oxford Institute of Statistics. In 1950 he returned to Austria and worked at the Austrian Institute of Economic Research until his retirement in 1978. He became an honorary professor of the University of Vienna in 1970, and in 1974 and 1975 he was a visiting professor at Stanford University. His publications include *Maturity and Stagnation in American Capitalism* (1952) and *Random Processes and the Growth of Firms* (1965).

Ilan Vertinsky is professor of policy analysis, management science, and resource ecology and chairman of the Division of Policy Analysis at the University of British Columbia. He received the B.A. in economics from the Hebrew University of Jerusalem and the Ph.D. in business administration from the University of California at Berkeley. He served on the faculties of Northwestern University and the Universidad del Valle before joining the University of British Columbia. He was also a Fellow of the International Institute of Management in Berlin. He has published extensively and his areas of interest are R&D and technological innovation policies, systems analysis, and resource management.

About the Editor

Devandra Sahal received the master's degree in nuclear physics from Helsinki University and a licentiate of technology in industrial engineering and production control as well as the Ph.D. in industrial management from Helsinki University of Technology. He started his career as a research scientist in industry in Helsinki, working in the areas of systems analysis and irradiation resistance of high polymers. In 1974, after four years in industry, he joined the faculty of the Systems Science Ph.D. Program at Portland State University, Oregon, where he developed and taught graduate courses in mathematical models in biology and ecology, research planning, general systems theory, and cybernetics. Since 1977 he has been a senior research fellow at the International Institute of Management in Berlin. He has served as a consultant on R&D management and national technology policy to several governmental and international institutions, including the World Bank and the United Nations. He is the author of more than forty scientific papers, which have appears in various journals including *AIIE Transactions, Cybernetica, Environment and Planning, General Systems Yearbook, IEEE Transactions on Engineering Management, International Journal of Systems Science, Pattern Recognition,* and *Technological Forecasting and Social Change.*